RIVERINE NEIGHBOURHOOD

Hydro-politics in South Asia

RIVERINE NEIGHBOURHOOD

Hydro-politics in South Asia

Uttam Kumar Sinha

idsa
INSTITUTE FOR DEFENCE STUDIES & ANALYSES
NEW DELHI

PENTAGON PRESS

Indigenous Historical Knowledge: Kautilya and His Vocabulary

Pradeep Kumar Gautam, Saurabh Mishra and Arvind Gupta (Editors)

First Published in 2016

ISBN 978-81-8274-866-8

Published by
PENTAGON PRESS
206, Peacock Lane, Shahpur Jat,
New Delhi-110049
Phones: 011-64706243, 26491568
Telefax: 011-26490600
email: rajan@pentagonpress.in
website: www.pentagonpress.in

Branch
Flat No.213, Athena-2,
Clover Acropolis,
Viman Nagar,
Pune-411014
Email: pentagonpresspune@gmail.com

In association with
Institute for Defence Studies and Analyses
No. 1, Development Enclave,
New Delhi-110010
Phone: +91-11-26717983
Website: www.idsa.in

Printed at Aegean Offset Printers, Greater Noida, U.P.

CONTENTS

PREFACE

In its conception the 'Riverine Neighbourhood' has a distinctive set of geographic perspectives that can be described as both the 'space of perception' and the 'space of myth'. While we easily comprehend issues that exist in time and relate it cogently to a historical context, we often tend to ignore the existence of issues in space and scale and the significance of geography. In a sense the book emphasises the natural/physical attribute of 'location' and how complex relationships evolve and interact by linking environmental-societal dynamics to economic and political systems. Rivers in South Asia as they criss-cross the political boundaries introduce interdependencies that can either reinforce or reduce differences.

The sub-continent is endowed with extraordinary natural and civilisational resources. The vast fertile stretches and the perennial rivers make it the land of abundance. JK Bajaj and MD Srinivas in their work *Restoring Abundance* cite the great American sociologist and demographer Kingsley Davis who gave a glowing account of the Indo-Gangetic plain as the greatest expanse of rich, tillable soil, and thus one of the world's greatest agricultural region and further said, "The geographical traits of the subcontinent are fabulous and their description requires unblushing superlatives..."[1]

The cultural development around rivers are shaped by the predictability and availability of resources. The Indian civilisation laid great emphasis on ensuring abundance of food and sharing the products among the living as a primary principle of righteous public functioning or the *dharma*. Our national song *Vande Matram* has the lines "*sujalam suphalam malayajasitalam sasyasamalam*" which translated is "rich with thy hurrying streams, bright with orchard gleams" – the land that is richly watered, richly fruited and richly harvested.

India is also a land of prolific myth making. The Hindu tradition, as it is well known, is famous for its mythologies and multitudes of gods and goddesses. But these have a distinct geographical detail to it describing the mountains, the river systems, and the holy places. Diana L Eck, Harvard Professor of Comparative Religion and Indian Studies, well researched work *India: A Sacred Geography* describes the distinction between how the historians of religion interpret ancient India and how cultural geographers study India. In so doing, she refers to the renowned geographer Bimala C Law, who said, "One finds it tedious to read the legendary history of *tîrthas* or holy places, but to a geographer it will never be a fruitless study."[2] For those further interested, Eck's book is a more expansive research of her earlier work *Banaras: The City of Lights*, which I was inspired to read in my numerous travels to Banaras Hindu University as an adjunct faculty at the Malaviya Centre for Peace Research.

The other word in the title of the book is 'Hydro-politics' or water politics. This is not a popular expression among water practitioners. In using hydro-politics, the book does not in any way negate hydro-cooperation rather the chapters argue that cooperation is hydro-politics. Since no water dispute, as history tells, has almost ever led to war, states have to ensure that sensible hydro-politics prevails so that the possibilities of water wars are unlikely in the future. Transboundary rivers link its riparians in a complex network of environmental, economic and security interdependencies.

Cooperation among South Asian riparians is undoubtedly high but that does not mean the absence of competing claims for water. Thus water will remain deeply political. Often water agreements are not always about water. History and hegemony play an important role in understanding the strategic interaction among riparian states and in the contextual framework under what circumstances politics interfere with cooperation or whether sharing of water acts as a neutralising factor in difficult political situations. Equally important is how history and interest influences riparian behaviour. I observed, as a member, in a number of CSCAP meetings (2012-2013) on water resources that China's initial reluctance and resistance to discuss water gradually gave way to more bilateral conversation and dialogue with the Lower Mekong Basin Countries on the Mekong. Even on the Brahmaputra, China has gradually moved to more openness on sharing hydrological data with India.

Rivers articulate land and landscape and are far older than the nation-state. Gopalkrishna Gandhi poignantly observes, "There was a time when rivers 'held' territories, defining their boundaries, giving their people a certain personality and both land and people, a certain ethos."[3] Hard politics today will proclaim that rivers belong to countries now. But rivers also flow and many a times crossing political boundaries and, therefore, it is not of exclusivity but of mobility and plurality or what can be described as a 'living complex system'.

In sum, *Riverine Neighbourhood: Hydro-politics in South Asia* is a study of how network of rivers give a sense of location and belonging –nationally and regionally and how critical it is to think about multi-level, inter-disciplinary cooperation on rivers that will include hydro-diplomacy at the inter-state and regional level; scientific knowledge and technical development of rivers and their flows; the cultural attributes; the economic significance; and transparency and hydrological data sharing.

The main objective of writing the book is to examine hydro-cooperation or, what is often described as, "sensible hydro-politics". Cooperation on water very simply put is a subset of the larger diplomacy in the region. What the book hopes to bring to the table is the importance of water issues across different levels of analysis and approaching the issue from a comprehensive perspective, rather than a statist understanding.

Uttam Kumar Sinha

NOTES

1. Cited in JK Bajaj and MD Srinivas, Restoring Abundance: Regeneration of Indian Agriculture to Ensure Food for All in Plenty, Shimla: Indian Institute for Advanced Studies, 2001
2. Cited in Diana Eck, *India: A Sacred Geography*, New York: Random House, 2012, p.45
3. Gopalkrishna Gandhi, "Hydro-wisdom must back hydro-diplomacy" in Ganesh Pangare (ed.) Hydro-diplomacy: Sharing Water Across Borders, New Delhi: Academic Foundation, 2014, p. 9.

ABBREVIATIONS

BCM	Billion Cubic Meter
BHEP	Baglihar Hydro Electric Project
COA	Court of Arbitration
CSE	Centre for Science and Environment
EIA	Environment Impact Assessment
EF	Environmental Flows
FFMP	Flood Forecasting Master Plan
GBM	Ganga-Brahmaputra-Meghna Basin
GGB	Greater Ganga Basin
GIS	Geographic Information System
GLOF	Glacial Lake Outburst Floods
GSBA	Ganga Strategic Basin Assessment
HKH	Hindu Kush-Himalayan region
ICA	International Court of Arbitration
ICIMOD	Centre for Integrated Mountain Development, Kathmandu
ILA	International Law Association
IPCC	Inter-Governmental Panel on Climate Change
IWT	Indus Waters Treaty
IUCN	The International Union for the Conservation of Nature
JCE	Joint Committee of Experts
JSCWR	Joint Committee on Water Resources
J & K	Jammu & Kashmir

JRC	Joint Rivers Commission
JSTC	Joint Standing Technical Committee
JTG	Joint Technical Group
KW	Kilowatt
MAF	Million Acre Feet
MGNREGA	Mahatma Gandhi National Rural Employment Guarantee Act
MoU	Memorandum of Understanding
MHA	Million Hectare
MRC	Mekong River Commission
NWAI	National Waterways Authority of India
WWAP	World Water Assessment Programme
MW	Megawatt
NASA	National Aeronautics and Space Administration
NGOs	Non Government Organisations
SAARC	South Asian Association for Regional Cooperation
SAWI	South Asia Water Initiative
TAR	Tibet Autonomous Region
TPE	Third Pole Environment
UN	United Nation
UNEP	United Nations Environment Programme

INTRODUCTION

WATERSCAPE: THE INESCAPABLE REALITY

'Water is life' is expressed in different languages, and through the practices and beliefs of many religions. In Hinduism, 'everything is water'; the Chinese express water as 'wou-ki' or the great uncertainty; and in the Koran 'God created all things from water'. It is nature's greatest irony that while water covers 70.9 per cent of the earth's surface, only 3 per cent is fresh or potable, of which 2 per cent is held in ice caps and glaciers. Of the 1 per cent, 30 per cent is ground water and a mere 0.3 per cent is found in lakes, ponds, rivers, streams, swamps, marshes and bogs, is non-frozen, salt-free, and accessible for human consumption.[1] It is this amount that truly matters in sizing future water challenges. More importantly, there is approximately the same amount of water on earth today as there was when it was formed.[2] Water is the only substance found naturally in three forms: solid, liquid and gas. Interestingly, there is more fresh water in the atmosphere than in all of the rivers combined, and if all of the water vapour in the earth's atmosphere fell at once and was distributed evenly, it would only cover the earth with about an inch of water.[3]

With population increase and consumption patterns, it is projected that two out of every three people on the planet will live in water-stressed conditions by the year 2030. In the last century, world population tripled and the use of water grew six times. By 2030, the demand for water will be 40 per cent more than it is currently and 50 per cent higher in the most rapidly developing countries that include India and China.[4] According to UN 2015 estimates, by 2030, world population is projected to reach 8.5 billion (low estimate) and 9.7 billion (medium estimate) by 2050 from the current level of 7.4 billion people. The bulk of the population increase will be in countries already experiencing water shortages.

Water, often described as the 'blue gold', is the lifeline for all life, and is often referred to as the 'commons' – resources that are meant for collective use and belonging to all rather than for individual ownership and exploitation. Walljasper says:

> Water commons means that water is no one's property; it rightfully belongs to all of humanity and to the earth itself. It is our duty to protect the quality and availability of water for everyone around the planet. This ethic should be the foundation of all decisions made about use of this life giving resource. Water is not a commodity to be sold or squandered or hoarded.[5]

Clearly, water as a non-substitute resource is the responsibility of both the state and the community to protect and nurture for purposes of current consumption as well as for future requirements, both in sufficient quality and quantity. There is now considerable emphasis on the 'commons' principle. Maude Barlow argues that: 'Every human activity now needs to be measured by its impact on water and the water commons,' and declares that 'It is a flagrant violation of human rights when only the rich have access to clean water'.[6] While not so common with South Asian countries, the idea of 'water commons' has been included in the Constitutions of South Africa, Uruguay, Ecuador and Bolivia as a human right, making it difficult for delivery systems to be sold into private hands.[7]

The UN Water website has recorded startling facts regarding water from various sources, including the World Bank, UNEP and the Oregon State University. Some of these facts are listed below:[8]

- 85 per cent of the world population lives in the driest half of the planet
- 783 million people do not have access to clean water.
- 6-8 million people die annually from the consequences of disasters and water-related diseases.
- Water cannot be seen in isolation – increasing agricultural output, for example, will substantially increase both water and energy consumption, leading to increased competition for water between water-using sectors.
- Agriculture accounts for 70 per cent of global freshwater withdrawals (up to 90 per cent in some fast-growing economies).
- The dietary shift from predominantly starch-based to meat and dairy will continue to impact water consumption over the next 30 years.

- The cost of adapting to the impacts of a 2°C rise in global average temperature could range from US$70 to US$100 billion per year between 2020 and 2050. Of this cost, between US$13.7 billion (drier scenario) and US$19.2 billion (wetter scenario) will be related to water, predominantly through water supply and flood management.
- Water is not confined to political borders. An estimated 148 states have international basins within their territory.
- There are 276 trans-boundary river basins in the world (64 trans-boundary river basins in Africa, 60 in Asia, 68 in Europe, 46 in North America and 38 in South America).
- 46 per cent of the globe's (terrestrial) surface is covered by trans-boundary river basins. 185 out of the 276 trans-boundary river basins, about two-thirds, are shared by two countries. 256 out of 276 are shared by 2, 3 or 4 countries (92.7 per cent), and 20 out of 276 are shared by 5 or more countries (7.2 per cent). A maximum of 18 countries share the Danube river basin.
- Russian Federation shares 30 trans-boundary river basins with riparian countries, Chile and United States 19, Argentina and China 18, Canada 15, Guinea 14, Guatemala 13, and France 10.
- European and North American populations consume a considerable amount of virtual water embedded in imported food and products. Each person in North America and Europe (excluding the former Soviet Union countries) consumes at least 3 m^3 per day of virtual water in imported food, compared to 1.4 m^3 per day in Asia and 1.1 m^3 per day in Africa.
- Land grabbing is another increasingly common phenomenon. Saudi Arabia, one of the Middle East's largest cereal growers, announced it would cut cereal production by 12 per cent a year to reduce the unsustainable use of groundwater. To protect its water and food security, the Saudi government issued incentives to Saudi corporations to lease large tracts of land in Africa for agricultural production.
- There are numerous examples where trans-boundary waters have proved to be a source of cooperation rather than conflict. Nearly 450 agreements on international waters were signed between 1820 and 2007. Over 90 international water agreements were drawn up to help manage shared water basins on the African continent.

These facts and figures tell a telling tale. First, water is indispensable and the ultimate renewable resource. Second, water is being severely impacted by global population increase and economic growth. Together, they are extracting and polluting it faster than it can be replenished. Third, the ever-expanding gap between demand (in terms of growing population and economy) and supply (in terms of availability) will potentially make water a contested issue, particularly in densely populated countries. Fourth, since disputes over water are inevitable because of the changes as described above, understanding the processes of resolution and framing new mechanisms and approaches become necessary.

Water is linked to every facet of human life. German philosopher Johann Goethe said that 'water is a friendly element for those who are familiar with it and know how best to treat it'. There are five broad characteristics of water that states should be *aware of* and *enlightened about* in order to frame well-rounded water policies. First, water is the largest shared resource in the world. Almost 90 per cent of water flows across or under political boundaries and therefore, requires considerable inter-state understanding. Second, water, as a trans-boundary resource, remains by nature unruly, and conforms to no unanimously accepted rules. Procedures do exist, based on principles and norms; but water-sharing arrangements among basin countries are reached ultimately through political equations. Third, despite this unruliness, states have shown a willingness to cooperate. Some of the world's most implacable enemies have negotiated water agreements, and others are in the process of doing so even while relations are strained. There are, thus, good practices and inter-state experiences available. Fourth, water issues within countries determine water issues between countries. Fifth, water is a resource that is extracted from the ecosystem and, therefore, the law of nature demands that it be returned to what is now referred to as the 'e-flow'.

The Run of Rivers

Rivers are the most visible form of fresh water. Rivers are ancient and older than civilisations – a 'mini cosmos' spawning history, tales, spirituality, and technological incursions. Flowing rivers are the largest renewable water resource as well as a crucible for both humans and aquatic ecosystem. Rivers also have a habit of moving on and on from their source from where they gush with gay abandon to their mouth where they quietly disappear into the surroundings. That journey is now being interrupted. Since the age of industrialisation, humans have increasingly exerted a pervasive influence on water resources. Rivers in particular have drawn humans to monumental engineering interventions such as dams and barrages – often as chest-thumping dominance and seldom as an enduring bond between man and nature.

As a result, we are swimming against the tide of unlearnt history and bad modern water policies. Most world rivers are in a state of crisis of degeneration and degradation. This is serious and frightening. Eighty per cent of the world population resides on the banks of rivers, and their well-being is threatened by increasing water pollution and flow fluctuation. Each river has its own distinct identity and profile; but when it comes to vulnerability and stresses, they are astonishingly similar. These include intensive water use for agriculture, unregulated industrial development and waste flow, and river habitat modification.

As rivers and water bodies become increasingly crucial to economic development and urbanisation, it will be important for national policies to look at the world as a network of rivers and as ecological systems interacting with their drainage basins as well as the atmosphere. Often policies lack sensitivity towards basic values like equity, transparency, accountability, sustainability, and the participation of the people in water policy processes. The beauty of successful river cleaning and rejuvenation is that it is not all about investing billions but equally about developing inter-disciplinary and inter-ministerial strategies. Thus, rivers have to be re-conceptualised as a relationship of history, culture and ecology and to be treated as endowments, to be sustained for future generations.

Rivers and Ecology

Various investigations and new sets of knowledge have awakened our understanding of rivers. Rivers are not being determined not merely in terms of flows and volumes but as integrated systems that include flood plains and riparian corridors. It is now widely understood that the e-flow concept is essential for sustainably managing water resources and meeting the long-term needs of people. The expanded knowledge about the various dimensions of rivers has also enlarged the benefits that can accrue from rivers. The World Wildlife Fund Report describes a free flowing river as one that:

> Flows undisturbed from its source to its mouth, either at the coast, an inland sea or at the confluence with a larger river, without encountering any dams, weirs or barrages and without being hemmed in by dykes or levees. In today's world such rivers, particularly those that run over long distances, are increasingly rare. In large river systems distinct stretches of rivers can retain characteristics of a free-flowing river, despite the presence of water infrastructure upstream or downstream of this stretch.[9]

The Report further describes ecological flow as the quantity, timing, and quality of water flows that are required to sustain a freshwater and estuarine ecosystem as well as the livelihood and the well-being of humans depends on these ecosystems.[10]

However, as is evident, rivers are increasingly being intervened with dams, barrages and diversions. A broad estimate suggests that more than half of the world's large rivers are dammed. Dams have enormous benefits; but they do change flow patterns which disturb the downstream quality of water as well as sediment movement and deposition. This subsequently affects fish and wildlife, and the livelihoods of people. In the context of South Asia, dams in rivers are important – indeed, the rivers in the region are amongst the most dammed. However, the challenges of maintaining 'comprehensive environmental flows' are enormous, with mistrust and apprehension being visible both upstream and downstream. In South Asia, the e-flow concept has not gained the desired relevance, and is contested with damming, embanking, and controlling rivers.

For about six decades of the 20th century, developed nations primarily focused on maximising flood protection and water management. By the 1970s, the negative ecological and economic impacts of water projects prompted scientists and developers to rethink and modify dam operations in order to maintain watershed principles and certain fish species. As a result, concepts like 'minimum flows' and 'in-stream flows,' came into the lexicon of water planners. At a global level, the 2007 Brisbane Declaration on Environmental Flows was endorsed by more than 750 practitioners from more than 50 countries.[11] Except for Bhutan and Maldives, each South Asian country had institutional representatives. The noted ones included the Kathmandu-based ICIMOD, as well as the International Water Management Institute (IWMI) offices in Nepal and Sri Lanka. University and research centres included the Bangladesh University of Engineering and Technology, the Cochin University of Science and Technology, the River

Research Centre and Agriculture University in India. The WWF offices in India and Pakistan also participated. According to the Declaration the key findings include the following:

- Freshwater ecosystems are the foundation of our social, cultural, and economic well-being.
- Freshwater ecosystems are seriously impaired and continue to degrade at alarming rates.
- Water flowing to the sea is not wasted.
- Flow alteration imperils freshwater and estuarine ecosystems.
- Environmental flow management provides the water flows needed to sustain freshwater and estuarine ecosystems in coexistence with agriculture, industry, and cities.
- Climate change intensifies the urgency.
- Progress has been made, but much more attention is needed.[12]

The Declaration announced an official pledge to work together to protect and restore the world's rivers and lakes. By 2010, many countries throughout the world have adopted environmental flow policies, although their implementation remains a challenge.

What then does Environmental Flows (e-flows) mean? Simply put, e-flows, as described by the IWMI mean 'an ecologically acceptable flow regime designed to maintain a river in an agreed or predetermined state'. Thus, e-flows are a balance between water resources development and river maintenance. They imbibe sustainable development principles. Another useful way of thinking about e-flows is that just as water is required for agriculture, industry, and domestic use, so also water is a requirement and a necessity for the river. Despite an easy understanding of the e-flows concept, difficulties arise in the estimating and determining what the value of e-flows is. This is because of the lack of understanding as well as the lack of available quantitative data regarding the relationship between river flows and the multiple components of river ecology. Noted Indian water expert, Ramaswamy Iyer also underlines the significance of distinguishing between in-stream flows for different purposes:

> Flows are needed for maintaining the river regime, making it possible for the river to purify itself, sustaining aquatic life and vegetation, recharging groundwater, supporting livelihoods, facilitating navigation, preserving

> estuarine conditions, preventing the incursion of salinity, and enabling the river to play its role in the cultural and spiritual lives of the people.[13]

As rivers meander from the source to the mouth, they are a dynamic combination of water, sediment, aquatic organisms, and riparian vegetation. The journey and the changes that occur are described in terms of a 'continuum' by ecologists.[14] One of the most fascinating features of the water is that it is neither created nor destroyed. It is simply a product of the dynamic hydrologic cycle, referred to as the 'mediated resource'.

River as Civilisation

Rivers are deeply connected to the growth and development of human civilisation. The rise and fall of civilisations was marked by the flow pattern of rivers, and their use for irrigation and transportation. Even in today's interconnected globalised world, a symbiotic relationship with rivers continues. At one level rivers are commercially exploited and are value addition (for example, dam building and power generation); at another level, they are an inseparable part of the identities, cultures, and religious perceptions of different peoples.

Rivers that cross national boundaries give a region a certain riverine identity, and a certain type of behaviour pattern. They also acquire geo-political characteristics resulting in rivalries and/or cooperation. At times, these identities transcend and connect; at others they divide political and territorial units. This raises an important question in the South Asian context: does a river represent a supra-identity which transcends national identities? In this context, the identity attached to rivers like the Indus, Ganga and Brahmaputra come to mind. All this is well documented and adequately expressed, and can help in ascertaining shared values and benefits across borders.

Water as Discourse: Symbolic Capital

Water has the same popular appeal as justice, freedom, equality, representation, and power. There is also something elemental or inherently wicked about water because searching for solutions to manage and cope with water issues creates a set of different problems that are political, emotive and divisive. As the most shared resource in the world, competition among various uses in trans-border river basins precipitating into conflict remains a concern. Yet, while the outlook for water is challenging, it has the ability to create breakthroughs, compelling different users to cooperate rather than allow confrontation to jeopardise water supplies.

According to Michel Foucault, 'Discourse constructs the topic. It defines and produces the objects of our knowledge. It governs the way that a topic can be meaningfully talked about and reasoned about.'[15] In positioning water issues as a discourse, it is essential to acknowledge the tenets of Didier Bigo's 'symbolic capital'.[16] Bigo contends that certain voices are inherently endowed with more weight than others due to the 'symbolic capital' inherent in them which is equivalent to positions of authority.[17] Bigo links this authority to knowledge, an idea which advances Foucault's power/knowledge equation.[18] Accordingly, a statement becomes 'power' when the audience takes the statement as 'true'. Various actors – political leaders, historians, the scientific community and the media – help in the 'mobilisation of knowledge resource' based on historical analyses, scientific evidence, and statistics. Those actors who are endowed with 'symbolic capital' and those who are concerned with the production of 'power/knowledge' form an important link in shaping the security discourse. Here, two examples primarily demonstrated to enlarge perception and locate water issues in the security logic are noteworthy.

In the early 1980s, Boutros-Boutros Ghali, the Egyptian Minister of State for Foreign Affairs, said, 'The next war in our region will be over the waters of the Nile.'[19] In 1991, a few months before being appointed as the Secretary General of the United Nations, he reiterated, 'the next war in the Middle East will be fought over water, not politics.'[20] Boutros Ghali was echoing the 'symbolic capital' of water, and expressing his recognition of the gravity of the situation in West Asia as deriving from historical analysis as well as his personal experience – both being a part of the 'power/knowledge' equation. For example, the Bible mentions that variations in water supply led to drought, and drove Jacob and his family to Egypt – an event that led to years of slavery and eventually to the consolidation of the Israelite tribe 400 years later.[21] Joshua directed his priests to stem the flow of the Jordan River with the 'power of the Ark of the Covenant' while he and his army marched across the dry riverbed to attack Jerico.[22] And, during World War I, as the Ottoman Empire crumbled, water resources became a critical factor in defining the territorial interests of the French, British, Arabs and Jews in West Asia.[23]

Moreover, through his own experiences as Egypt's Minister of Foreign Affairs from 1977 to 1991, Boutros Ghali had seen that emotions could run high over the sharing of the region's most precious resource. Thus, when President Anwar Sadat offered the waters of the Nile to Israel in a bid to open discussions about the West Bank and Gaza, there was public outrage in Egypt and beyond, with upstream countries protesting that the waters of the Nile were not President Sadat's to distribute at will. Thereon, 'water wars' as dramatic alliteration became fairly common and even became the title of an article written by Joyce Starr.[24] In 1995, World Bank Vice-President Ismail Serageldin made a much-quoted prediction about the future of war, 'If the wars of this century were fought over oil, the wars of the next century will be fought over water.'[25]

In his Independence Day speech in 2004, India's Prime Minister Manmohan Singh highlighted the importance of water. This was yet another example of the 'symbolic capital' of water. He identified water as one of the *saat sutras* requiring special attention. The challenge outlined by him was one of managing water resources as well as ensuring people's participation in water management and conservation:

> Water is a national resource, and we have to take an integrated view of our country's water resources, our needs and our policies, as well as our water utilisation practices. We need to ensure the equitable use of scarce water resources...I urge you and all our political leaders to take a national and holistic view of the challenge of managing our water resources.[26]

Earlier, in his address to the nation on 24 June 2004, Manmohan Singh had said: 'Water has emerged as a critical and contentious issue across the country...The government will reverse the neglect of public investment in irrigation, addressing the specific problems of each river basin, in an environment and people friendly manner'.[27]

For Prime Minister Narendra Modi and his government also, water is central to governance. He has expressed *sujalam sufalam* (water for prosperity) as a *yojana* (plan). River rejuvenation is critical to his development plans. The earlier Ministry of Water Resources has now been expanded to include River Development and Ganga Rejuvenation. Likewise, Drinking Water and Sanitation, which was a department under the Ministry of Rural Development, is now a full-fledged ministry, with a cabinet rank minister heading it.

What then does 'symbolic capital' explain? First, it helps in understanding securitisation as a performative act, or as Ole Waever says a 'speech act'.[28] Speeches and statements thus become a reference point. According to Barry Buzan, security is a practice 'quality actors inject into issues by securitising them, which means to stage them on the political arena...and then to have them accepted by a sufficient audience to sanction extraordinary defensive moves.'[29] Second, the speech act is not merely political rhetoric but signifies 'specific rhetorical structure'[30] in which the securitised issue is presented as an issue of supreme priority – a movement from 'low politics' to 'high politics'.

Water in a New Security Framework

The fundamentals of security/insecurity – who is secure from whom or what, when, where and how – will always remain embedded in the security discourse,[31] which to use a significant line, 'Every concept like security...has a story to tell; a story of their own coming to presence'.[32] In 1983, in his thought-provoking article titled 'Redefining Security',[33] Richard Ullman introduced a new approach to understanding international security by incorporating non-military considerations such as environmental dangers, disease, hunger, natural disasters, and population growth. Ullman wrote, 'the non-military tasks are likely to grow ever more difficult to accomplish and dangerous to neglect.'[34]

In understanding climate change and its impact on national security, it is important to separate the 'how' question (how will climate change lead to conflict? And how it has come into the mainstream of discussions on security?) from the 'where' question (where will such conflict occur?). This section analyses the 'how' question. The 'where' question is dealt in subsequent sections.

In the Cold War period, non-traditional security issues found little or no space in the security matrix of the time. The security framework was built by being able to maintain the capability to defeat or deter aggression. Military strength was the key. With the disintegration of the Soviet Union and the end of Cold War, non-military issues in a post-Cold War of peace dividends, institution building, and new approaches to resolving conflict got considerable attention. Since non-traditional security issues cut across borders, the dominant Cold War security themes of 'territoriality' and 'impermeability' held little ground. Owing to the existential nature of such issues, active reconciliation rather than the Cold War theme of mutual recrimination gained currency. Non-traditional security issues by their very nature challenged the Cold War notion of security based on unilateral solutions and the advocacy of military actions. With the demise of the Cold War and the growing scientific evidence of the impact of climate change on food, water and energy, there has been a systemic attempt to redefine security, moving away from the exclusive focus on the traditional notion of a state's ability to protect itself. Instead, there are increasing interest centres which are broadening the definition of security to include 'newer' and more non-traditional threats that can undermine political stability, undercut economic productivity, or erode levels of human well-being. With climate change, food, energy and water, national security in the 21st century has seen a radical departure from viewing the world 'as it is' to understanding the world 'as it ought to be' – that is, in coexistence with nature.

In understanding the new dimensions of national security, it is important to consider that the impacts of water resource affects not only conflicts as diverse as war, terrorism, or diplomatic and trade disputes but also conflict within states. Two points need to be noted:

1) There is no single causal factor to water-induced conflict. It varies from case to case. In some it may be a major factor; in others it may be a minor one.
2) Water-induced conflict is not an entity in itself but part of complex pathways to conflict that involves political, strategic, economic and territorial factors.

Thus, water issues are compelling the redefinition of national security instead of merely defending it. In many ways, water issues challenge the state-centric proprietorship of security. Traditional security frameworks are antithetical to climate change issues for the simple reason that their impact does not respect state borders, and therefore limits/prevents states from taking unilateral action. Secondly, in the traditional understanding of security, the protection of territorial integrity is primarily based on the threat from an enemy 'other'. In the case of challenges to water availability, the threat comes from the imbalances in the ecosystem and the enemy is scattered: the state, people and the corporations. The people are enemies in pursuit of a better quality of life (consumption patterns); corporations are enemies in pursuit of profit (business lobbies); and the state is an enemy focusing on the symptoms rather than the causes (the here-and-now policies). Thirdly, in the traditional security approach, the participation and contribution of actors in enhancing the understanding of security is limited, whereas mapping water threats and seeking remedies to prevent them requires broad-based participation. In other words, there is greater participation of the 'epistemic community' (a trans-national network of diplomats, experts, academia, and civil society) in the national and international decision-making process for the effective formulation of water policies.

However, there has been marked resistance from both the traditional security community as well as the development community to link security with water issues. The traditional security community argues that water issues are primarily welfare and development concerns, and that the state has always been oriented towards protecting water supply. The development community feels that positioning water issues into the security ambit will only reinforce the state-centric apparatus.

The primary security concern over the impact of water is the potential for violence, conflict, or military action as a result of – and in response to – widespread water scarcity. The debate, however, revolves on the approach to the water threats: whether they should be co-opted into the state-centric security framework or whether the approach should be one of securing the ecosystem in which water is both a security referent and a security goal. The latter approach argues that it is better to focus separately on the components of water issues – that is, water (hydro) diplomacy, water (hydro) cooperation, basin approach, watershed management, and environment flows. The 'securing the ecosystem' approach suggests trade-offs – for example, cutting down on arms expenditure for forestation programmes, soil conservation, water efficiency, and demand-side water management. Viewing the impact of water scarcity from the 'state-centric security framework' has its pitfalls. For example, water can become another tactic used by developed countries to impose their values on developing countries thus infringing their sovereignty – that is, a tool of hegemonic power. Moreover, water security rhetoric encourages thinking that could lead nations to undertake military interventions in the name of protecting 'global' resources.

Given the stress on waters both in terms of quantity and quality, it is not unlikely that states will try to maximise water resources and convert them into assets to augment their power. Downstream countries which are highly dependent on river waters for their well-being will be motivated to seize such a vital resource from their neighbours to the point of even being aggressive. Studies and assessments reveal that conflict and turmoil related to water will be as much internal as external – that is, affect bilateral relations in South Asia. Thus, water management and the reallocation water resources will be a major national security concern.

Contesting the Cold War exclusivity of security issues has enabled a discourse that deconstructs the realist theories of the state being the unitary actor. In the post-Cold War period, non-traditional aspects of security have been subject to a high degree of scholarly debate and research. The period has been a fruitful one for thinking about a broader agenda in security issues, both conceptually and in policy terms. While critics of 'broadening the security ambit' dismiss it outright (by threatening 'to destroy its intellectual coherence and make it more difficult to devise solutions to any of these important problems'),[35] its proponents, in a true Hobbesian sense, reason security through its multiple meanings.[36]

In 1993, in order to come to terms and make sense of the rapidity of change in the international system, the Copenhagen School led by Ole Waever, Barry Buzan, and others worked on the shift in the referent object being the state to the referent object being society in general – that is, the so-called 'securitisation' of international relations.[37] Thus, the Copenhagen School (CoS) provided theoretical grounds for the conceptualisation of non-traditional security. Along with the military, the political, the economic and the societal, the environment became one of the five different sectors of security that interact and interconnect.[38] Thus, security became a mode of reasoning that required protecting the referent object. The security discourse is now increasingly focusing on the dynamics of 'securitisation/de-securitisation' as well as 'politicisation'. It is argued that the securitisation of an issue advances the 'friend/enemy construction' while de-securitisation is emphasised at the societal level on ethical considerations.[39] Buzan, however, contends that securitisation is an extreme version of politicisation. Politicisation makes an issue relevant and involves responsibility; on the other hand, securitisation involves the urgency of a threat, which legitimises actions outside the normal bounds of political procedure.[40]

The scholarship that followed suggests a relationship between the environment, especially resource scarcity, and violent conflict – that is, the Scarcity Model.[41] However, establishing a causal link has proven elusive.[42] Thomas Homer-Dixon's work underlines the relationship between the environment and conflict as an interactive and complex one, and that environmental stresses and strains can be important contributors to conflict even if causally distant.[43] In particular, he posits that environmental scarcity has insidious and cumulative social impacts, such as population movement, economic decline, and the weakening of states, which can contribute to sub-national violence.[44] These impacts can provide challenger groups with opportunities for action against a state that has been gradually eroded by civil war, corruption, economic mismanagement, rapid population growth, or deteriorating renewable resources.

The scarcity model often referred to as the resource-deficiency thesis has its critics, particularly on the question of how tension and the resultant stress from scarcity can become transmuted into armed violence in the form of large-scale conflict.[45] Clearly, the model still requires rigorous tracing of the relationship between resource scarcity (as a key determinant) and its impact on war-making and war prevention.[46] Nonetheless, Homer-Dixon reinforces his argument by saying that 'theorists have usually focused on the possibility of inter-state conflict over resources. We are claiming that because environmental scarcities are worsening we can expect an increase in the frequency of conflicts with an environmental component'[47] Drawing upon the security debate, particularly the 'scarcity model', water resource thus becomes both an existential and immediate threat, and an important determinant in understanding the stresses in the new international system.[48]

Three factors contribute to water resource being a scarcity threat: depletion and degradation; increased demand, and uneven distribution. Those concerned with water crises and their future are divided into two schools. One, led by Aaron Wolf, indicates that water, as a source of conflict is more likely to occur within countries than between them. This school focuses on water as a source of cooperation, and an impetus for scientists and political leaders to use modern science and advanced technology to create new solutions and suitable alternatives.[49] The Wolf School also looks into the history, scope, and design of international water treaties.[50] The other, led by Peter Glieck, argues that water scarcity as a source of conflict will be increasingly inter-state in nature, and examines water-related conflicts. Glieck, however, makes it very clear that 'water resources have rarely been the sole cause of conflict', and should be viewed as a 'function of the relationships among social, political, and economic factors, including economic development.'[51] The Glieck School also evaluates the role of water as a tool and weapon (both political and military) of conflicts caused by other factors.

Thus, security practitioners need to take into account water issues as part of their arsenal of tools. They should explore two primary questions: What role do water issues play in stimulating international conflict and cooperation? Are conflicts over water sharing likely to be more 'within' (intra-state) or 'between' states (interstate)? The Wolf-Glieck divide in terms of scope and focus is of obvious policy importance, particularly since threats emanating from water scarcity feature regularly as policy reports.

The biggest challenge for decision makers in the 21st century is how to reassess the nature of threats. Today's threats and challenges—described as 'shared risks and vulnerabilities'—are less predictable and more multi-dimensional. This has a significant bearing on the understanding of security. The 'broadening and deepening' of security polices from 'dead end choices' to a balance of social, economic and environmental polices has become a necessity.[52]

Various United Nations (UN) reports indicate that, by 2050, the lack of water rather than lack of arable land will constrain food production. The water–food link is crucial for sustained development and economic growth. This inextricable linkage in politically difficult regions makes water a source of economic and social instability. Moreover, the underlying fact that sovereignty over the resource is not determined by formal legal principles exacerbates the complexity. Managing freshwater resources will fundamentally require a change in the ways in which states and societies think about water. While it is essential to formulate water policies that are holistic, participatory, and ecologically sound, the political significance of water – especially in regions where water availability is increasingly being challenged by growing populations – cannot be overlooked. A stable supply of water will be crucial for political stability in densely populated regions.

With water concerns growing increasingly urgent, the global community will benefit from the lessons learned as well as best practices in water dispute resolutions and approaches to water management. Based on some of the familiarities of water – such as being the largest shared resource with trans-boundary characteristics; a mediated resource with watershed principles; an unruly resource seeking political interventions, and as a catalyst for cooperation – this book raises the following key questions:

- Where are the most pressing areas of tension over water allocation and utilisation?
- Are these areas covered by water-sharing agreements or dispute-resolution mechanisms?
- What negotiations are in progress, and what do they intend to achieve?
- Which mechanisms have proven effective in other parts of the world? Which should be avoided?
- How can hydrological data be best standardised and shared?
- How is climate change affecting water security and the prospects for sharing or joint management?
- What role can regional organisations play in setting standards or fostering cooperation?
- What steps should regional governments pursue in resolving water management issues with their neighbours?

South Asia Water Resources

Regions are now increasingly viewed as hydrological units, and no region with shared water is exempt from water-related controversies and disputes. A stable supply of water is critical for regional stability, and any economic and geopolitical forecasting has to factor the water resource. The Indian subcontinent/South Asia – home to about 34 per cent of Asia's population which is about 1/6th of world population – has about 4 per cent of world's annual renewable water resources that flows through several river basins.[53] Almost 95 per cent of water in South Asia is consumed by the agriculture sector as compared to the world's average of 70 per cent. Except for Nepal and Bhutan, the per capita water availability is lower than the world average Given the uneven endowment and development of water resources in South Asia, the issues and challenges are large, diverse, and complex.

South Asia is a region of multiple crises where water is connected to many of the challenges of development, security, and economic growth. Described as a crowded, hot, hungry, and a fast evaporating region, South Asia occupies about 5 per cent of the world's land mass and has about 4 per cent of world's annual renewable water resources that flows through several river basins. Almost 90 per cent of water is consumed by the agricultural sector. By 2025 the region will be home to about 25 per cent of the world's population. About three-quarters of South Asia's population live in rural areas, and one-third live in extreme poverty (less than a dollar a day). Societal challenges are compounded by the fact that the region is highly vulnerable to climate change, particularly the retreat of Himalayan glaciers and the changing precipitation which affects the flow pattern of the perennial rivers such as the Indus, Ganga, Sutlej and Brahmaputra. These great rivers, in turn, are the lifeline of tens of millions of people in Bangladesh, Bhutan, India, Nepal, and Pakistan.

The Himalayan Watershed

Many Himalayan rivers are intimately tied up with the issue of territory as the rivers enter areas where there is contestation over the demarcation of borders. For example, the Indus flows through parts of Kashmir that is labelled 'disputed' territory. For Pakistan, laying claim to Kashmir in effect means claiming the waters of the Indus river system. Similarly, the Brahmaputra is linked with the Sino-Indian border dispute in the eastern Himalayas, where China claims the territory of Arunachal Pradesh where the Brahmaputra enters India. Since most of the rivers originate from, flow through, and drain into territorially defined boundaries, it will not be easy to ignore the competitive nature of water as well as the significance of the Himalayan watershed from where the shared rivers originate. Indeed the latter may well be described as the hydrological faultline. South Asian states have developed along river systems that are intricately connected from the source (the glaciers in the mountains) to the mouth (the deltas). This interdependence has been less understood when policymakers focus on borders and territorial disputes.

That said, it is also important not to ignore the fact that all the trans-boundary rivers in South Asia cascade down from the towering heights of the Himalayas, thus resulting in an enormous hydro-potential, particularly in Jammu and Kashmir, Nepal and Bhutan. The various assessments of climate change on the glaciers suggest that there is going to be, in the short-to-medium time, an increase in melt-flow, resulting in increased flow and flooding. The construction of facilities to store this excess water and release it during dry periods bedevils planners, given the temptation to generate benefits on the one hand and give rise to other dangerous spin-offs on the other. This is particularly relevant to the Indus basin as glaciers roughly account for 45 per cent of flow. Beyond the economics of water management, including the need for dams and water storage facilities for economic development, there is the political reality of fear among lower riparian, especially over such structures. Clearly, the hydrology of the region is not only tied up with economic development but also with security and misperception.

Hydro-politics

With water assuming centrality, and increasingly becoming both a bilateral and regional agenda, South Asia is now a 'hydro-political security complex' in which states are simultaneously part 'owners' and part 'users' of rivers. This framework has opened up various levels of analysis on how riparian states behave (hydro-behaviour), upstream-downstream contestation (hydro-competition), prior use issues, and clashes of priorities. Given that states are rational egoists interested in maintaining relative capabilities, water has now acquired a political sharpness and the attributes of power.

Water relations can never be permanently settled, the reason being that river flows are not constant. The flows in turn are determined by seasonal variations and usage, particularly those that are non-consumptive in nature. Also, interventions and diversions on rivers impact flow. Political relations can easily be impacted by changes in the quantitative and qualitative nature of the river. Varied interpretations of the use of river water have resulted in claims and counter-claims.[54]

There are, however, accepted legal norms of 'equitable utilisation', 'no-harm rule' and 'restricted sovereignty' that riparian states work through, and frame negotiations and treaties to overcome such differing positions. These have been reflected in many of the water treaties in South Asia. Given India's riparian linkage, whether upstream or downstream, and given its diplomatic investment in a number of treaties with its riparian neighbours, hydro-diplomacy will be a vital component of its neighbourhood policy. A greater complexity has been added now that China is a hydro-heavyweight in South Asia's hydrography.

NOTES

1. US Environment Protection Agency, at http://water.epa.gov/learn/kids/drinkingwater/water_trivia_facts.cfm
2. See, 'Water Facts and Trivia', at http://www.lenntech.com/water-trivia-facts.htm

3. US Environment Protection Agency, at http://water.epa.gov/learn/kids/drinkingwater/water_trivia_facts.cfm
4. Mckinsey Report, 'Charting our Water Future', November 2009, at http://www.mckinsey.com/App_Media/Reports/Water/Charting_Our_Water_Future_Exec%20Summary_001.pdf
5. Jay Walljasper, at http://otherworldsarepossible.org/claiming-protecting-water. Cited in *Blues Beyond Boundaries: Transboundary Water Commons*, Action Aid Report, Bhubaneswar: Natural Resource Knowledge Activist Hub, 2015, p.1.
6. Maude Barlow, 'Water as a Commons: Only Fundamental Change Can Save Us', *The Wealth of the Commons; A World beyond Market &State*, Levellers Press, 2012.
7. *Blues Beyond Boundaries: Transboundary Water Commons*, Action Aid Report, Bhubaneswar: Natural Resource Knowledge Activist Hub, 2015, p.1.
8. http://www.unwater.org/water-cooperation-2013/water-cooperation/facts-and-figures/en/
9. 'Free-flowing Rivers: Economic Luxury or Ecological Necessity?', WWF Report, 2006.
10. Ibid., at http://www.eflownet.org/viewinfo.cfm?linkcategoryid=4&linkid=64&siteid=1&FuseAction=display
11. The Brisbane Declaration, Environmental Flows Conference, held in Brisbane, Australia, 3–6 September, 2007. http://www.eflownet.org/download_ documents/brisbane-declaration-english.pdf. Cited in Action Aid Report, *Blues Beyond Boundaries: Transboundary Water Commons*, Bhubaneswar: Natural Resource Knowledge Activist Hub, 2015, p.2.
12. Ibid.
13. Ramaswamy Iyer, 'The Notion of Environmental Flows: A Caution', NIE/IWMI Workshop on Environmental Flows, New Delhi, 23-24 March 2005.
14. Stanford, J.A. and Ward, J.V., 'An Ecosystem Perspective of Alluvial Rivers: Connectivity and the Hyporheic Corridor', *Journal of the North American Benthological Society*, 12, 1993, pp. 48-60.
15. Cited in Alec McHoul and Wendy Grace, *A Foucault Primer: Discourse, Power and the Subject*, Victoria: Melbourne University Press, 1995, p. 67.
16. The term 'Symbolic Capital' was coined by Pierre Bourdieu. It is to be understood to include the following: '...the concept of capital should be seen not only in economic terms, but also as applicable to a range of other resources such as knowledge and status, and these kinds of capital can be converted into each other.' Didier Bigo's 'symbolic capital' can be found in his study on internal security at the European level, including the roles of the police and the military.
17. See, Claudia Aradau, 'Migration: The Spiral of (In) Security', E-journal, at http://venus.ci.uw.edu.pl/~rubikon/forum/claudia1.htm
18. Ibid. Foucault's 'power/knowledge' is a reciprocal, mutually reinforcing relation between the production and circulation of knowledge and subsequently the control and exercise of administrative power. See Alec McHoul and Wendy Grace, op. cit. no. 4, p. 82.
19. Quoted in Daniel Hill, *Rivers of Eden: The Struggle for Water and the Quest for Peace in the Middle East,* New York: Oxford University Press, 1994, p. 66.
20. *The New York Times*, 25 May 1991.
21. Genesis, Chapter 41.
22. Joshua, Chapter 4.
23. Ironically, the Sykes-Picot Agreement did not take water into consideration. Other factors such as the location of holy places, rail and oil lines, and political alliance took precedence.

See, David Fromkin, *A Peace to End All Peace: The Fall of the Ottoman Empire and the Creation of the Modern Middle East*, New York: Avon, 1989, p. 99.

24. Joyce R. Starr, 'Water Wars', *Foreign Policy*, 82 (Spring), 1991, pp. 17-36. Starr is generally given the credit for coining the term 'water wars'.
25. Quoted in Joseph Nevins, 'Resource Conflicts in a New World Order', *Geopolitics*, 9(1), March 2004, p. 258.
26. Prime Minister Manmohan Singh's Address to the Nation on Independence Day, 2004, *Strategic Digest*, 34 (9), September 2004, p. 1272.
27. *Strategic Digest*, 34(7), July 2004, p. 978.
28. 'One can view security as that which in language theory is called a speech act: it is not interesting as a sign referring to something more real – it is the utterance itself in itself that is the act: by saying it something is done.' Ole Waever, *Concepts of Security*, Copenhagen: University of Copenhagen Press, 1997, p. 221.
29. Barry Buzan, Ole Waever, and Jaap de Wilde, *Security: A New Framework for Analysis*, London: Lynne Rienner Publishers, 1998, p. 204.
30. Ibid., pp. 14, 204. Buzan says: 'for the analyst to grasp this act, the task is not to assess some objective threats that 'really' endanger some object to be defended or secured; rather, it is to understand the processes of constructing a shared understanding of what is to be considered and collectively responded to as a threat'. Also, in 'Rethinking Security after the Cold War", *Cooperation and Conflict*, 32(1), 1997, p. 14, Buzan explains the 'rhetorical structure' based on existential threats to the referent object, protecting through exceptional measures the threatened referent object, which justifies and legitimises the actions.
31. James Der Derian, 'The Value of Security: Hobbes, Marx, Nietzsche, and Boudrillard', in David Campbell and Michael Dillon (eds.) *The Political Subject of Violence*, Oxford: Basil Blackwell, 1991, p. 97.
32. Michael Dillon, 'Security, Philosophy and Politics', in Mike Featherstone et al., *Global Modernities*, London: Sage, 1995, p. 158. Dillon's statement has an uncanny resonance to Buzan's performative act.
33. Richard Ullman, 'Redefining Security', *International Security*, 8 (1), 1983, p. 153; Jessica Tuchman Mathews' article of the same title appeared in *Foreign Affairs*, 68 (2), 1989, pp. 162-177.
34. Ibid., p. 153.
35. Stephen M. Walt contends that any attempt to widen the discourse on security is always resisted. See, 'The Renaissance of Security Studies', *International Studies Quarterly*, 35, 1991, p. 212.
36. In Thomas Hobbes' view, nothing in life can be achieved without security. If environmental degradation and resource depletion can be a source of future conflicts, then they are worthy of being in the security framework. See George H. Sabine and Thomas L. Thorson, *A History of Political Theory*, Mohan Primlani for Oxford and IBH Publishing, New Delhi, 1973, pp. 427-29. See also, Bertrand Russell, 'Hobbes's Leviathan', in *History of Western Philosophy*, Routledge, London, 1995, pp. 531-41.
37. In the Copenhagen School, referent objects ranged from the state to collective identities to the survival of individual species and even the habitat. See Barry Buzan, Ole Waever and Jaap de Wilde, no.29, pp. 22-23; see also, Ole Waever, Barry Buzan, Morten Kelstrup and Pierre Lemaitre, *Identity, Migration and the New Security Agenda in Europe*, Pinter Publications, London, 1993. Buzan coined the term 'Societal Security'; Waever preferred the term 'Identity Security'.

38. Barry Buzan, *People, States and Fear: An Agenda for International Security Studies in the Post-Cold War Era*, London: Harvester Wheatsheaf, 1991, p. 17.
    See also, E-journal, at http://venus.cr.uw.edu.pl/~rubikon/forum/claudia2htm.Aradau endorses de-securitisation techniques while arguing that de-securitisation can be equally unethical by reducing individual freedom and liberty.
39. Claudia Aradau, "Beyond Good and Evil: Ethics and Securitisation/Desecuritisation Techniques", E-journal, at http://venus.cr.uw.edu.pl/~rubikon/forum/claudia2htm.Aradau endorses de-securitisation techniques while arguing that de-securitisation can be equally unethical by reducing individual freedom and liberty.
40. Barry Buzan, Ole Waever, and Jaap de Wilde, n.29, p. 23-24.
41. The scarcity-conflict model is fast becoming conventional wisdom in foreign policy and population and environment circles. It has been structured by the likes of Stephan Libiszewski and Homer-Dixon, and popularised and sensationalised by writers like Michael Renner, 'Ending Violent Conflict', *Worldwatch* Paper No.146, 1999, and Robert Kaplan, 'The Coming Anarchy', *Atlantic Monthly*, February 1994, pp. 44-76. Kaplan proclaimed the environment as the most important national security issue of the 21st century.
42. Critics of a de-centred security argue that because proving the causal link has been difficult, the environment should be left to 'protection' rather than 'securitisation'. Daniel Deudney argues that turning the environment into an object of national security risks under-mining the positive forms of global environmental thinking and cooperation that have been emerging in recent years. See 'Environment and Security: Muddled Thinking', *Bulletin of the Atomic Scientist*, April 1991, pp. 22-28.
43. Thomas F Homer-Dixon, 'Strategies for Studying Causation in Complex Ecological Political Systems', Occasional Paper, American Association for Advancement of Science, Washington D.C., June 1995, p. 6.
44. Thomas F Homer-Dixon, 'Environmental Scarcities and Violent Conflict: Evidence form Cases', *International Security*, 19 (1), 1994, p. 5.
45. For the scarcity debate, see 'Environmental Security and Violent Conflict: A Debate', at http://www.ics.st.edu/PROGRAMS/DIS/ECS/report2/debate.htm
46. The scarcity model through comprehensive modelling exercises has been undertaken at the University of Toronto under Homer-Dixon.
47. 'Environmental Security and Violent Conflict: A Debate', at www.ics.st.edu/PROGRAMS/DIS/ECS/report2/debate.htm, p. 17.
48. Read, Michael T. Klare, 'The New Geography of Conflict', *Foreign Affairs*, May/ June 2001.
49. Aaron Wolf, 'Conflict and Cooperation along International Waterways', *Water Policy*, 1(2), 1998, pp. 252-65; Alsom Sandra Postel and Aaron Wolf. 'Dehydrating Conflict', *Foreign Policy*, September/October 2001, pp. 60-67. Wolf coordinates the Transboundary Freshwater Dispute Database, Oregon University, which includes a computer database of over 400 water-related treaties, negotiating notes and background material on 14 case-studies of conflict resolution, news files on cases of acute water-related conflict, and assessments of indigenous/traditional methods of water conflict resolution.
50. See, Aaron Wolf's 'Transboundary Freshwater Dispute Database', at http://www.transboundarywaters.orst.edu/
51. Peter Glieck, 'Water and Conflict: Fresh Water Resources and International Security,' *International Studies*, 1, 1993, p. 92. See also 'Water, War and Peace in Middle East',

Environment, 36, 1994. Glieck heads the Pacific Institute for Studies in Development, Environment and Security, Oakland, California, at http://www.pacinst.org/

52. Maria Trombetta, 'Environment Security and Climate Change: Analysing the Discourse', *Cambridge Review of International Affairs*, 21(4), 2008, p. 585
53. 'Freshwater Under Threat: South Asia', UNEP Report, 2008
54. See, Uttam Kumar Sinha, "India's Water Woes", *The Sunday Pioneer*, March 16, 2014 at http://www.dailypioneer.com/sunday-edition/agenda/cover-story/indias-water-woes.html

1

South Asia's Water Security

This chapter presents a broad understanding of water security. It examines trans-boundary rivers and the controversies and contestation surrounding the availability and distribution of rivers in South Asia. Fundamentally, the interdependent relationships rivers impose make them essentially interstate in nature. This may be about meeting domestic water needs, or controlling floods, or generating hydro-electricity. While the political dimensions are critical in determining water relations between states, an emphasis on 'Hydrodiplomacy' under the framework of sustainable development and the principles of 'efficiency, equality, equivalence and equity' cannot be separated from bilateral and regional dynamics. This opens up scope for wider stakeholder participation as well as opportunities for integrating science and different and varied techniques. However, using water as a diplomatic tool will need to overcome several challenges: these include competitive politics; power asymmetry; hegemonic analysis; misunderstanding; and lack of communication.

Rivalry over water is age-old, and is actually built into our language. It is said that Lord Buddha's first public act was to actually arbitrate a dispute over water. In fact, the word rival derives from the Latin *rivalis*, originally meaning 'person using the same stream as another'.[1] The phrase to 'sell someone down the river' means to betray someone.[2] Many analysts make a case that a number of conflicts in the future will be over water;[3] others take a contrary position arguing that 'water wars' are loosely conceptualised in the resource scarcity thesis.[4] Moreover, historical evidence records great understanding and participation towards settling water disputes. While the likelihood of tension and conflict emanating from the consumption and distribution pattern of water resources cannot be underestimated, historically, such resources have been used more as the means or rationalisation of conflict than seen as its cause. Thomas Naff, a noted hydrologist, describes hydro-political problems as being paradoxical in nature, exhibiting a tendency to encourage negotiations where other problems degenerate into conflict.

Hydrologic reality appreciates the rationality of water (it cannot be, in absolute terms, controlled and commanded), and strongly dictates that water (hydro) diplomacy and water (hydro) cooperation is the best possible means of optimising trans-boundary river waters. Even the 1997 UN Convention on the Law of the Non-Navigational Uses of International Water Courses drawn from the earlier Helsinki Rules, in spirit states the need for the 'equitable utilisation of water resources' and 'meeting vital human needs'.[5] The ability to cooperate on trans-boundary rivers limits the 'geopoliticisation' and 'instrumentalisation' of water. The 'preciousness' and 'possession' in geopolitical mechanics renders water a contested commodity. The 'geopoliticisation of water' associates with it the 'instrumentalisation of water', and therefore, the common usage of the term 'water wars'. Thus, water becomes a resource of contention and conflict which is generally reduced to the question of need and want, and what the cost of procurement would be in economic, political, or military terms.

The 'geopoliticisation' and 'instrumentalisation' of water leads to a 'dangerisation process' in which the 'preciousness' of water is to be 'possessed'. The water war hypothesis is a product of this framework. Crucially, it helps to focus on key issues such as river basin management and equable distribution, as well as on the assessment and monitoring of the potential impact of climate change on the water resources of any regions.There are essentially two types of water in question: big water and small water. The water that we drink and that is used for domestic consumption – which is about 10 per cent of the water needed – is small water. The big water concerns the production of food. 90 per cent of water consumed by society is needed to raise its food and underpin livelihood. It is this 90 per cent that is crucial to states, and central to the treaties which are usually about water for irrigation, dams, and hydroelectricity production.

Since the sources for water tensions will be more diverse—stemming from a combination of internal and external considerations and of the broader fallout of environmental change – any hydro-cooperation framework will have to be harmonised simultaneously at the national, regional, and local levels. More importantly, this emphasises that peaceful co-existence through cooperation is rational and more effective than a conflictive approach to water allocation. This analysis does not make alarming predictions of wars over water; however, it does highlight the stresses and strains between riparian states in spite of the momentum towards water cooperation. By asking the question 'who gets how much water, and why?' one can foresee the varying intensities of conflict and the power symmetry/asymmetry amongst the riparians.

Another significant feature of international river basins points to the fact that there are a number of bilateral accords governing multilateral basins, which are described as 'fragmented governance'.[6] The predominance of bilateral arrangements is viewed as a 'by-product of asymmetric power within the basin'.[7] To be successful in river basins, multilateral accords require a greater degree of power parity amongst the riparian states. This is often difficult to achieve. While analysing 'asymmetric power' and 'power parity' in the river basin, many experts have frequently examined the nature of hegemony. The hegemonic state is defined as one which has a substantial concentration of material capabilities in the river system. Such an assessment also distinguishes a hegemon who is generous and benevolent from one that is aggressive or predatory. The hegemonic analysis or the 'hegemonic reference point' has frequently featured in the study of trans-boundary rivers, and how states behave (unilaterally) in relation to their hydrological profile or riparian position.[8]

The distribution of power in a river treaty formation is also much debated and contested. Much of the debate centers on river basins like the Nile, Euphrates and Tigris, and the Mekong. Some analyses suggest that an upstream hegemon is less likely to sign a water agreement, while a downstream power, in all probability, will impose a treaty on its weaker upstream riparian to secure its interests.[9] There is a prevailing observation that power asymmetry seen as a difference in wealth rather than military strength has a greater effect on treaty formation.[10] This means that a prosperous country in a river basin that values interdependent economic gains will help facilitate cooperation irrespective of whether it is lower or upper riparian.[11] Other findings based on upstream and downstream power distribution point towards more bilateral accords on multilateral basins.[12]

Some scholars have examined how certain river disputes get resolved, and the factors/conditions that facilitate an agreement.[13] This functional analysis attracts policy-making. Broadly, the findings indicate that riparian agreements are made with regard to basins where water is of high value, and where disputes over water are frequent. Discords are an important initiator of accords. The prevailing discords in such river basins involve present and current practices rather than future problems. Advancing hydro-diplomacy in the regional context would mean understanding the factors facilitating cooperation in various parts of the world. There have been numerous water agreements and treaties (nearly 3600) taking into account basin dynamics, hydrological knowledge as well as societal values. While some have been fundamentally weak, many have had the resilience to withstand the tension. Mapping the international basins and understanding the strengths and weaknesses of the existing treaties or water arrangements as well as charting the best practices will have a positive role in influencing river treaty formation in the future, as also help strengthen hydro-diplomacy.

The Importance of Water Regimes

As Robert Keohane and Joseph Nye argue, the international system is increasingly interdependent, and the behavior of states and the ensuing politics affect the level of interdependence.[14] Any efforts towards cooperative arrangements are usually referred to as international regimes. Rivers, as stated, are trans-boundary and interdependent, and hence prospects of basin cooperation would enhance the prospect of hydro-solidarity amongst the riparian states.

Water regimes have distinct characteristics, such as a set of rules to reduce conflict caused either by use, pollution, or the diversion of a water resource.[15] Some water regimes are general in nature: for example, the 1992 Agenda-21 which relates to the protection of the quality and supply of freshwater resources. Similarly, the 1997 UN Convention on the Law of the Non-Navigational Uses of International Water Courses established general principles for the use of trans-boundary water resources. More recently, in 2008, the UN General Assembly adopted the resolution on the 'Law of Transboundary Aquifers,' encouraging states to cooperate on the sustainable use of cross-boundary aquifers. Other regimes are more specific, and are intended for the particular resolution of conflict in a watercourse, as for example the Rhine regime which dealt with issues of pollution from chemicals and chlorine. A water regime helps to bring sensitive and troublesome issues up front and facilitate settlements of disputes. The UN has often sought the involvement of a third party, which is trusted by contesting parties, to help the process of cooperation through not only water allocation provisions but also through the monitoring and enforcement mechanism. International Relations (IR) scholars have given detailed explanations on how regimes come about, and who shapes them.

From a realist perspective, regimes are created by powerful hegemons to serve their interests. The robustness of the regime is reflected by the power of the hegemon; 'as the power of the hegemon declines, the regime also weakens'.[16] For the neo-liberals, however, norms are of more value than the hegemon because norms prevail and guide states to evaluate actions and reactions and also to temper behavior. Regime formation, as some others view it, is also a direct consequence of a crisis or shock or disturbance. The role of expert communities in the formation of a regime is important. The epistemic community/multidisciplinary experts in both the formation and transformation of the water regime can greatly influence policy, and can also lead to a general convergence of policies internationally. Each water regime formation on a river basin can have good and bad practices, and each can learn from the other and can serve as a baseline for inter-state water relations. The importance of water regimes cannot be dismissed by hydro-diplomacy. By promoting inter-state cooperation and mitigating conflict, an evolved water regime decreases water insecurity. For example, the Jordan River Basin between Israel and Jordan is a good example of how regime formation and cooperation have been successful despite competitive and conflictual politics. The proponents of regime formation also cite the Indus Waters Treaty and the Ganga Treaty, where the convergence of values and the cooperation within the regime have been institutionalised, making it increasingly difficult to reverse them, or end cooperation.

Hereon, this chapter will outline how treaty formation, power distribution, and hegemony analysis play out in South Asia, which is now being observed from a hydrological perspective which includes the Himalayan watershed and the Tibetan glaciers from where many of the Asian rivers originate.

The Dynamics of River Treaties

Almost half of the earth's land surface lies within international river basins, and constitutes a significant share of the world's available supply of fresh water.[17] There are certain obvious but often overlooked attributes to river waters that shape the argument of water security. One, that it traverses borders and hence is a trans-boundary issue. Two, it is the most widely shared resource on the planet. The fate of a riparian nation that shares a river basin is inextricably tied to the river. Quite literally, 'A River Runs Through It'. River waters define inter-state relations, and the structures of treaties will be increasingly tested. There are more than 260 river basins that are shared by two or more states, and there are 145 treaties in existence today.[18] More than 45 per cent of the world's population lives in internationally shared river basins. It is commonly acknowledged that, in spite of non-binding international law and rules for managing river water basins, treaties serve as the best management tool. A large number of riparian treaties reinforce an argument that river waters are a 'catalyst' for cooperation even among hostile states rather than an 'inducement' for conflict. In the past 50 years, there have been only 37 cross-border disputes which have involved violence, while numerous initiatives on water-sharing have been negotiated and signed.

River treaties reflect a measurement of cooperation and offer states a structure to coordinate actions. There are various propositions that govern river treaties. For example, scholars have investigated that freshwater scarcity motivates cooperation particularly those that are bilateral in nature.[19] A critique to such investigation is that while scarcity provides the main impetus for cooperation, it does not necessarily mean that a straight forward, linear relationship will develop between the two riparian states. Other externalities than just scarcity determine the strength and trajectory of water treaties. Building upon the relationship between scarcity and cooperation, some experts stress on an 'inverted U-shaped curve'.[20] This curve suggests that states in a river basin are less likely to cooperate in a situation where water per capita is either very low or very high.[21] Between the extremes of high and low is the moderate level of water availability which is likely to evince cooperation as it will require smaller mitigation costs.

The basis for any river water treaty is to continuously find an equitable approach for meeting vital human needs. Water treaties, particularly in regions where scarcity is high, are also a barometer to gauge state behaviour and the political climate. They raise a few interesting (not necessarily tautological) observations: to what level does a changing political climate effect existing treaties? Does the signing of river water treaties lead to more cooperative ventures between the riparians concerned and thereby enhance the overall peace environment in the region? Is the negotiation process preceding the signing of a treaty a final solution? Or, is it only a provision that temporarily conceals the claims and counter-claims and the real and perceived fears of the riparians (particularly the lower riparian)? Do 'the real and perceived fears' lead to non-compliance of the treaty with an overriding 'militarised' approach in which the 'possession' of water is determined unilaterally? And finally, what are the linkages associated with trans-boundary waters?

River water treaties are time-specific, and are a translation of a political will at a particular time and cannot be viewed in terms of finality. As the lives and livelihood of people exponentially grow around the river basin, so does the demand for and consumption of water. The efficacy of treaties between the riparians will always be tested, if not completely severed or abrogated. Signs of open hostility of an armed nature directly related to water have seldom been witnessed after the treaty has been signed. It proves the point that water, in spite of the challenges and difficulties is, at the end of the day, solvable whether through cooperation and treaties, or technology and investment. This turns the water war thesis on its head. In fact, the West Asian region was closer to water wars in the 1960s than one could possibly imagine today. And now, even the Palestinians and the Israelis are negotiating on water issues under the Joint Water Committee of the Interim Agreement.

Based on some of the readings of the characteristics of the existing 145 water treaties (but with the main focus being on water treaties in water-stressed areas), the following important observations can be noted:[22]

- River water treaties are both 'rights-based' and 'needs-based'. The former is dominant during the course of negotiations. In the post-treaty period, issues are more 'needs-based'. That is why most treaties in water-stressed areas have to 'cope' with changing ground realities.
- Both the upper and the lower riparians have stakes in the continuation of a treaty, with non-water linkages playing an important role. The upper riparian in most cases is a strong military power, which leaves the lower riparian to seek non-military ways, and linkages become a crucial component of a treaty. Such linkages or tactics to 'enlarge

the pie' include, according to the databank, 'capital, land, technical support, [and] political concessions'.

- Water-stress regions also witness difficult political environments; however, in spite of the animosity and suspicion, cooperative ventures on trans-boundary water resources have been a dominant theme. A third-party role or the inclusion of the United Nations is critical: first, to help initiate a cooperative framework on the shared water basin; and then, to financially assist in the projects. A third party role is an attractive mechanism for lower riparians to enter into a treaty arrangement as it offers space for third-party intervention/adjudication of disputes.
- While 'rights-based' to 'needs-based' define river water treaties, the role of a conflict resolution mechanism in the existing treaties has been far less sophisticated, and would need strengthening. With the growing importance of 'river watershed management' (not just the harnessing of the 'surface water'), and with easy availability of new monitoring technology, an entirely new set of enforcement mechanisms can be structured and infused into the treaty with lasting value. The rationalising of water cannot be bereft of politics.

The depoliticisation discourse might have merits; but one cannot avoid the controversies that accompany water decisions. To make sense of these controversies, the interface between politics and water has to be carefully assessed and analysed in order to create favourable ground for sound decisions on water reforms and water management. It is important to understand how power structures influence water policies, at both the local and national level (intra-state) as well as the regional level (inter-state). The water war trend needs to be reversed through national water management plans, and also by the incorporation of water in the framework of regional peace and cooperation. Thus, improving trans-boundary watercourses cooperation is of utmost importance. Cooperation on water will help strengthen regional economic development as well as cultural preservation.

South Asia: A Riverine Region

Trans-boundary river basins[23] are a prominent feature of the South Asian physical landscape, cutting across political boundaries and are, therefore, of paramount importance to the region's geo-political stability. Trans-boundary rivers physically link upstream and downstream users, and their uses offer ample opportunity for harnessing development benefits. While they can serve as a cornerstone for cooperation, they can equally be a potential entry for tension and strife. Adding to the complexity, trans-boundary rivers are neither seen exclusively as a 'public good' (defined as non-rival and non-excludable) or a 'private good' (defined as rival and excludable).[24] While indeed such an amorphous definitional demarcation subjects trans-boundary rivers to various interpretations, it has however in the context of the Sub-continent been dealt with fairly well through bilateral treaties based on norms, and an understanding of water sharing and, distribution. That said, there is however, a general view to perceive it as 'collective goods' or 'common pool resources'. Rivers have many uses. Some are 'consumptive' in nature, and some 'non-consumptive'. The non-consumptive uses, such as navigation and hydroelectricity generation, are less problem-generating than consumptive uses such as drinking water and water for irrigation. Because river uses are subjective in terms of the where, what, and how they are being used, a water-tight riparian treaty seldom emerges.

The two crucial river basins in the Sub-continent include the Indus in the west and the Ganga-Brahamaputra-Meghna (GBM) in the east. The two basins make for an interesting analysis of the driving forces behind water demand, on the political obstacles that stand in the way of river cooperation, and the behaviour of the riparians. Any future assessments of South Asia cannot ignore the hydrology of the region, and its interplay between the physical, the political, and the economic. The hydrological profile of the two basins includes the vast water resources of Tibet. This, inescapably, brings into observation China's water concern and its riparian approach.

Water in South Asia is a very sensitive and highly political issue. Rivers are a primary source of freshwater in the region, and are intimately tied up with the issue of territory. For example, the water of the Indus is crucially linked with territorial issue in Kashmir. Similarly, the Brahmaputra is linked with border issues with China, particularly in Arunachal Pradesh. Essentially, water dynamics in these two river systems have resolved around 'sharing the waters' and 'sharing the benefits'. It is also very important to recognise that all of the trans-boundary rivers in South Asia come out of the great heights of the Himalayas and, therefore, there is enormous hydro-potential, particularly in Nepal. However, the mountain kingdom has only developed

3 per cent of the potential. Comparatively, Pakistan has 10 per cent and India 25 per cent. The various assessments of climate change on the glaciers suggest that there is going to be an increase in melt-flow, resulting in more flooding. Building up capacity to store this excess water and release it in dry periods is enormous. The contribution of snow and glacial melt varies from the eastern part of the Himalaya to the western. The glaciers contribute to about 10 per cent of the flow in the Brahmaputra and the Ganga, while it accounts for almost 45 per cent in the Indus. Beyond the water dynamics as explained above is the political reality of the region – the fact that it is the least integrated region in the world. Thus, water across the region is not only tied with economic development but also with security.

While the rivers link countries together, they also bitterly divide them. Riparian issues are ultimately a political issue. Politics is about power, influence, resource allocation, and policy implementation. Politics is also about managing relationships and trade-offs between states. In many ways, water management will be crucial to conflict management in the region. Water being indispensable is an emotional issue that can become a cornerstone for confidence building and a potential entry point for peace. However, an appropriate discourse that shifts away from 'water war' to 'water peace' needs to be developed.

Of the many uses of rivers, it is the non-consumptive uses such as navigation and hydroelectricity generation which are less controversial than the consumptive uses such as drinking water and water for irrigation. With mounting population pressures and the need to achieve developmental goals, disputes and grievances arise over the use of and control over rivers. Structures like dams and barrages create upper-lower riparian tensions that have the potential to lead to conflict. The numerous bilateral treaties are often hostage to prevailing political animosity. Given the pressures of water demand, resource nationalism will increasingly dominate the hydrological contours of South Asia.

In *realpolitik*, the preciousness of water often translates into possessiveness and, at times, even to resource aggressiveness. 'Water insecurity' can be expressed in terms of availability, reliability, and quality, while 'water security' can be described as insecurity arising from control of headwaters. The insecurity-security dynamics can be explained through upper-lower riparian relations, and further examined through the claims of 'absolute territorial sovereignty' versus 'absolute territorial integrity'.

Water simultaneously brings hope and despair in South Asia. While conflict over water seems increasingly unlikely, disputes have been frequent, and will remain so. These are, of course, being settled, or resolution is being attempted at different levels. The difficulties of water issues are those that relate to availability and equity which combine with political factors to create impediments to resolving water disputes. Often ignored is the fact that political will has many a time helped settle disputes under the existing water agreements. Water treaties and agreements often bring in a paradox to South Asia. While the region is largely seen as unsettled, fractious, and the least integrated on the water front it has shown maturity to settle water disputes and political wisdom to cooperate. In the last 55 years, since the Indus Waters Treaty was signed, some generalisation from water experiences in South Asia can be drawn:

- The Indus Waters Treaty of 1960 is scrupulously upheld by India and Pakistan though all other bilateral agreements hang under cloud or remain ineffective.
- Despite the preparation and consensus, the Mahakali Agreement of February 1996 between India and Nepal was jarred with mistrust and misgivings. Similarly, the Ganga Water Agreement of December 1996 between India and Bangladesh still has its hitches. On the other hand, India's water agreements with Bhutan proceed smoothly and to mutual advantage.
- Politics is the determinant of settlement or non-settlement, not water *per se*.
- More often than not, decisions have been inordinately delayed in implementation resulting in suspicion. Moreover, ground realties and conditions change very quickly, making the earlier decision meaningless. The late V G Verghese would often describe the paradox as a matter of lacking good water governance.
- Water disputes are largely the product of change, territorial (as resulting from Partition), as an exponential function with population pressures, changes of usage, etc.

South Asia through a Hydrological Lens

Water is now being increasingly viewed as an issue of urgency in South Asia. With its rising population, increasing urbanisation, and unchecked poverty, the Subcontinent has added enormous pressure to existing water sources. With no proportional increase in availability, water challenges seem imminent. The trans-boundary nature of water, as seen through the rivers that crisscross the South Asian states, makes it intensely political and contentious while simultaneously creating opportunities for hydro-cooperation. The shared nature of these challenges requires both macro and micro-level collaborations, such as integrated water management efforts between governments.

South Asian states are part of the river systems intricately connected from the source (the glaciers in the mountains) to the mouth (the delta). This interdependence has been less understood and often de-prioritized over conventional and territorial disputes. This is ironical in many ways since some of the major trans-boundary rivers crossing boundaries in South Asia (for example the Indus and the Brahmaputra), are in areas that remain politically contentious. The more South Asia comes to be viewed as an exponential function – that is, increase in population leading to a greater demand for food, larger claims for areas of cropland, and greater volumes of water – a better hydrological understanding of the region will emerge. Significantly, planning any water resource utilisation policy will have to take into account the assessment of the impact of climate change in terms of seasonal flow and extreme events. In both direct and indirect ways, climate change is related to water as is evidenced through floods, drought, and glacial melt.

From a hydrological perspective, China cannot be removed from the South Asian regional configuration. China is not member of the SAARC, a political grouping; but it acquired observer status along with Japan, South Korea, and the USA in 2009. Increasingly, and as India's neighbouring countries like Bangladesh, Pakistan and Nepal would like, China is making its presence felt in South Asia and, in the process, competing directly with India which considers the region to be its sphere of influence. From a hydrological position, India is a lower riparian vis-à-vis China, and an upper riparian vis-à-vis Pakistan and Bangladesh. An emphasis that has not been correctly articulated is the fact that India is also middle riparian, and has concerns over water uses with China as well as the responsibility of sharing waters with its lower riparian neighbours. China's hydrological position, on the other hand, is one of upstream supremacy. India's middle riparian position increases both its dependency on the head waters of rivers sources such as Indus, Sutlej and Brahmaputra which originate in the Tibetan plateau, as also the pressure of sharing with its down riparian Pakistan and Bangladesh.

'Asymmetric power' and 'power parity' govern the river basins in South Asia – that is, the Indus and the GBM. The hegemonic reference point cannot be ignored, with China and India as the states with substantial concentrations of material capabilities in the river system. Importantly, such assessment differentiates a hegemon that is generous and benevolent from one that is aggressive or predatory. Since China is the ultimate upstream country (with no formal river sharing agreement or treaty with its neighbours) and India is middle riparian (with a number of water treaties with its neighbouring countries), the hegemonic analysis would suggest that China exhibits a negative hegemonic role on the waters as compared to India. In fact, the hegemonic analysis would place India in the category of a generous hegemon.

Another important feature which each individual state in South Asia has to consider is to integrate and harmonise external water policies with internal water resource management. Such an approach would require treating river systems – particularly the GBM and the Indus – in a holistic way, and reorienting hydro-diplomacy on a multilateral basis rather than just a bilateral format. This would entail a shift from 'sharing waters' to 'sharing benefits'. Ecological considerations should be the overarching perspective. This would easily allow a far greater understanding of the impact of climate change on water resources. In the past, the dominant perspective was engineering and economics; now the emphasis should be on ecology and climate change. Keeping the principle of just and wise use of water, sensible riparian policies in South Asia can be framed. And, this also includes the effective participation of China.

South Asian states will have to juggle competing and conflicting food-energy-water (FEW) concerns, yielding a set of difficult consequences. A 'perfect storm' of food-energy-water shortages by 2030 has already been predicted.[25] These sets of critical drivers will present difficult-to-manage outcomes, and will reinforce each other as never before. First, as population grows, competition for food, energy, and water will correspondingly increase. Increasing demand for food grains will claim larger areas of cropland and greater volumes of irrigation water. Second, with the risks that climate change attaches, FEW will be subject to various stresses and strains. Clearly, for

South Asian countries, food security cannot be achieved without water security. India, for example, feeds 17 per cent of the world's population but has only 4 per cent of water.

South Asia is home to about 34 per cent of Asia's population (1/6th of world's population), and has about 4 per cent of world's annual renewable water resources that flow through several river basins.[26] Almost 95 per cent of water in the region is consumed by the agriculture sector as compared to the world's average of 70 per cent. Except for Nepal and Bhutan, the per capita water availability is falling below the world average. It is projected that the per capita water availability in India is rapidly declining. For the year 2025, at a projected population of 1.3 billion, the water availability will be 1341 cubic meter/person/year.[27] Of significant importance is the fact that planning any water resource utilisation policy will have to take into account the assessment of the impact of climate change in terms of seasonal flow and extreme events. In both direct and indirect ways, climate change is related to water as is evident through floods, drought, and glacial melt.

Over the next 20 years, rising concerns about the effects of climate change will take greater precedence over any physical changes associated with climate change. In all likelihood, perceptions of a rapidly changing ecosystem will prompt nations to take unilateral actions to secure resources and territorial sovereignty. Any willingness to engage in greater river basin cooperation will depend on a number of factors – such as the behavior of other competing countries, the economic viability, and other interests that states are reluctant to either compromise or concede.

A Tale of Two Trans-boundary River Basins

The Indus Basin is an important geophysical feature of the Indian subcontinent. The Indus, together with the Chenab, Ravi, Sutlej, Jhelum, Beas, and the extinct Sarasvati, forms the 'Sapta Sindhu' delta in the Sindh province of Pakistan. The Indus originates in the Tibetan plateau in the vicinity of Lake Mansarovar, runs through Ladakh, Jammu and Kashmir, Gilgit-Baltistan and Pakistan before it merges into the Arabian Sea. The total length of the river is 3,200 km. The river's estimated annual flow is approximately 207 billion m^3.

Table 1: Indus Basin

Total Basin Area	1170838 km²
Annual Available waters	224 billion metric³
Country	*Basin Area (Km²)*
Pakistan	632,954
India	374,887
China	86,432
Afghanistan	76,542

Source: *Freshwater Under Threat: South Asia*, UNEP Report, 2008 http://www.reliefweb.int/rw/lib.nsf/db900sid/ASAZ-7NZJEX/$file/unep_Dec2008.pdf?openelemen. *P26*

The GBM flows through the northern, eastern and northeastern parts of India. The river system is as much a blessing as a curse because of its water potential and destructive reality. The basin covers an area of about 1.75 million km2, with an estimated population of approximately 535 million (75.8 per cent in India; 20 per cent in Bangladesh; 3.5 per cent in Nepal; 0.2 per cent in Bhutan; and 0.5 per cent in China). The basin has huge development opportunities; but it is also home to the largest concentration of poor in the world. The majority of the population (approximately 10 per cent of the global population) subsists on agriculture.

Table 2: Ganga-Brahamaputra-Meghna Basin

Total Basin Area	1745000 km²
Annual Available Waters	2,025 billion metric m³
Country	*Basin Area (Km²)*
India	1,105,000 (62.9 per cent)
China	326,000 (19.1 per cent)
Nepal	140,000 (8.0 per cent)
Bangladesh	129,000 (7.4 per cent)
Bhutan	45,000 (2.6 per cent)

Source: *Freshwater Under Threat: South Asia,* UNEP Report, 2008 http://www.reliefweb.int/rw/lib.nsf/db900sid/ASAZ-7NZJEX/$file/unep_Dec2008.pdf?openelemen, p. 26.

Riparian Relations

Riparian nations are those 'across which or along which a river flows, have legal rights to use the water of river'.[28] Simple as it may seem, varied interpretations have resulted in claims and counter-claims between riparian countries. Though disregarded by the international water laws, upper riparian nations essentially base their claims on 'absolute territorial sovereignty' – that is, the right to use the water of rivers the way they want regardless of lower riparian concerns. The lower riparian nations, on the other hand, base their claims on 'absolute territorial integrity', claiming that upper riparians can do nothing that affects the quantity and quality of water flowing downstream. Clearly, both the claims are extremely incompatible. To overcome such differing positions, riparian states work through the more accepted legal norms of 'equitable utilisation', 'no-harm rule' as well as 'restricted sovereignty', and frame negotiations and treaties accordingly.[29] However, all these norms in state politics and power politics are rendered meaningless. It is a vague notion that nations are entitled to a 'reasonable share of water'.[30] Keeping in mind that there is no legal-binding international treaty on water sharing, how will riparian politics play out in the region? The future of conflict or cooperation in the Indus and GBM basins will revolve around the water 'insecurity-security' dynamics of the basin states. Below is an overview:

1. China and India will be critical players in the hydro-politics of the region. India is simultaneously an upper, middle and lower riparian. China's hydrological position, on the other hand, is one of complete upper riparian supremacy. Unlike China, India has a high dependency on the head waters of the sources of rivers such as Indus, Sutlej and Brahmaputra which originate in the Tibetan plateau. 'Water insecurity' in China relates to the disproportionate availability of waters, the majority of which are in the south (the Tibet Autonomous Region), with the north and west excessively water stressed. Also, pollution is a big worry, with a vast majority of lakes and many of its largest rivers unsuitable for human use.[31] In terms of 'water security', China has no threat. It is probably, along with Turkey, the world's most independent riparian country.[32] This position gives it enormous flexibility in shaping larger political equations with its neighbours. On the other hand India, given its longstanding commitment to bilateral river treaties (which China has none)[33] has to assiduously balance the anxiety and concerns of its lower riparians (Pakistan and Bangladesh) without compromising its own water requirements. For India, both 'water insecurity' (internal water management strategy and its neighbourhood approach) and 'water security' vis-à-vis China is high.

2. An evaluation of the other actors in the two Trans-boundary River basins, particularly Pakistan and Bangladesh, suggest that 'water insecurity' in terms of per capita availability, reliability, and quality is severe. But the bigger fear is in terms of 'water security'. Pakistan is dependent on the sources of rivers outside its boundaries, but fears India more as an upper riparian than China which it has allowed building-operating-transfer of river projects in Gilgit-Baltistan (in Pakistan-Occupied-Kashmir) for hydroelectricity. For Bangladesh, being the lowest riparian in the GBM, its dependency is extreme vis-à-vis India, with 94 per cent of the waters originating beyond its borders including the 54 rivers, rivulets, and streams. With both countries, water will increasingly become a political and emotional issue, and a critical driver of the larger politics with India. However, there are treaties and mechanisms to address the issue – like the 1960 Indus Water Treaty now in its 50th year, and the 1996 Ganga Treaty which is valid until 2026. Unlike the other basin countries, Nepal and Bhutan have high per capita water availability, and therefore, 'water insecurity' is low. The issue, however, is how best to share the benefits of water with India. This has failed to achieve the required success in Nepal's case as political complexities, the implementation of treaties, and perceptions have undermined the enormous potential for water cooperation. However, with Bhutan water cooperation has been a win-win.

A snapshot of the riparian behaviour suggests that no country will like to be dependent on either India or China having control over an indispensable resource. The inherent lack of trust between the two countries accentuates this. For the basin states other than China, being water independent is not an option. For India, water issues will be far more political and strategic vis-à-vis Pakistan and China. Though politics cannot be discounted from India's water relations with Nepal and Bangladesh, there is considerable scope to overcome and break political deadlocks through sensible water sharing arrangements and resource development. In building hydro scenarios (2030), three important variables have been identified: a) the water resources of Tibet; b) China's rise; and c) the Indus Water Treaty between India and Pakistan. A brief description of the variables follows below before the scenarios are drawn.

Himalayan Hydrology

It is being fast established that Himalayan hydrology will be one of the critical frontlines in the global battle against climate change and water scarcity. The Himalayan mountain system is of crucial importance to the river system of South Asia not only in terms of influencing the monsoon but also in terms of the glaciers which are the source of many of the great rivers in Asia. Geologists often regard all the rivers, including those originating from Tibet, collectively as the 'circum-Himalayan rivers'.[34] As studies indicate, the impact of global warming and climate change will gradually shrink glaciers, resulting in the decrease of water runoff in the long-term. In the short-term, earlier water runoff from glaciers when combined with seasonal rains can result in flood conditions.

Over the next 20 years, perceptions of a rapidly changing ecosystem may prompt nations to take unilateral actions to secure resources and territorial sovereignty. Any willingness to engage in greater river basin cooperation will depend on a number of factors, such as the behaviour of other competing countries, the economic viability, and other interests that states are reluctant to either compromise or concede.

The risks and uncertainties over the impact of climate change on water resources are potentially high in many South Asian countries. For example, given its location and geography, Bangladesh is extremely vulnerable to any variations in water flow. Being the lowest of the riparian states, it shares 54 rivers with India. Geographically speaking, Bangladesh is in a double trap: on the one hand, river flows make it increasingly water dependent; on the other, it is witnessing sea-level rise. According to a modelling study, the mean global temperatures for Bangladesh may rise by 1.5 to 1.8 degree centigrade by 2050, and correspondingly sea levels may rise by about 30 cm accompanied by an increase in annual rainfall.[35] For India, the middle riparian, decreased snow cover will affect the flows in the Indus, the Sutlej, the Ganga and the Brahamaputra – all originating from Tibet. 70 per cent of the summer flow of the Ganga comes from the melt-water and, thus, can potentially impact the agriculture sector. India's National Communications (NATCOM) in 2004 has projected a decline in wheat production by 4-5 million tonnes, with even a 1 degree centigrade rise in temperature. Like Bangladesh, Pakistan is a lower riparian, and is vulnerable to access of clean water. The western Himalayan glaciers act as reservoirs that release water into the rivers that feed the plains in Pakistan. The glacial retreat is increasing the flow, and the recent devastating floods in Pakistan in July-August 2010 are a stark reminder of the perils of climate change. In the next decade, erratic rainfall combined with glacial melt will exacerbate the already serious problems of flooding and draining. After the glacial has receded, it is projected that there will be a 30-40 per cent reduction of flow in the Indus basin, critically impacting food production.[36]

Tibet: the Third Pole

The Tibetan Plateau, regarded as the 'Third Pole', is a storehouse of bountiful freshwater captured in its massive glaciers, large lakes, and cascading waterfalls. After founding the People's Republic of China in 1949, the new communist regime sought reunification with Tibet and invaded it in 1950. Following its occupation, it made extensive modifications in the borders of the provinces of Tibet. Amdo was made the new Qinghai province. U Tsang and eastern Kham were designated as Tibet Autonomous Region (TAR) of China, and the remaining parts merged into its provinces of Sechuan, Yunnan and Gansu.[37] With an area spanning 470,000 sq km, TAR is the second largest province of China after Xinjiang.

The plateau serves as the headwaters of many of Asia's largest rivers, including the Yellow, Yangtse, Mekong, Salween, Brahmaputra, Indus and Sutlej. It is estimated that the net hydrological flows in Tibet total 627 cubic km per year.[38] This is roughly 6 per cent of Asia's annual run-off, and 34 per cent of India's total river water resources. The availability of fresh water is 104,500 cubic meters per year, making Tibet the fourth largest waterhouse in the world after Iceland, New Zealand and Canada.[39] Various figures estimate that about 1.5 to 2 billion people in South and Southeast Asia are dependent on the watersheds of these rivers whose sources are in Tibet. These vast water resources are also vulnerable to climate change, and a host of serious environmental challenges that require global attention. The impact of global warming on the glaciers is also alarming with glaciologists suggesting that the glaciers in the plateau are receding at a rate faster than anywhere else in the world.

Tibet's water resources raise difficult and contesting questions. Should China alone be the stakeholder to the fate of the waters in Tibet? China has rampantly exploited all the rivers from the Tibetan Plateau except for the Nujiang River which flows through Yunnan and enters Myanmar, where it is known as the Salween.[40] The North-South diversion plans on the Yarlung

Zangbo (Brahamaputra in India), the only other river untouched, are afoot, causing fears and apprehensions for India.[41] With historical disagreement over the territory, Tibet's unresolved political status will be of direct consequence to ways to sustainably manage water resources and ensure that the natural hydrological flow of its rivers are not disturbed by Chinese artificial diversion plans. Lower riparian pressure and international attention to defining vital resource as 'commons' would be significant in preserving and sharing the waters of Tibet. While such re-definition is politically sensitive, as it clashes with national jurisdiction, it nonetheless, merits attention, keeping in mind the future water requirements of South and Southeast Asia and its dependency on water resources in Tibet.

NOTES

1. Wordreference.com. Also see, Merriam Dictionary.
2. Ibid.
3. For example, see recent work by Steven Solomon, *In Water: The Epic Struggle for Wealth Power and Civilisation*, New York: Harper Collins, 2010; Brahma Chellaney, *Water: Asia's New Battleground,* Washington, DC: Georgetown University Press, 2011; Rustom Irani 'Water Wars', *New Statesman and Society* 149 (4), 1991, pp. 24–25; 'Streams of Blood, or Streams of Peace', *The Economist,* 387, 3-9 May 2008 (available at http://www.economist.com/node/11293778) and Russell, Ben, and Nigel Morris, 'Armed Forces are Put on Standby to Tackle Threat of Wars over Water', *The Independent,* 28 February 2006 (available at http://www.independent.co.uk/environment/armed-forces-are-put-on-standby-to-tackle-threat-of-wars-over-water-467974.html).
4. Aaron Wolf and Jesse Hamner, 'Trends in Transboundary Water Disputes and Dispute Resolution', in *Environment and Security: Discourses and Practices* (ed.) Miriam Lowi, and Brian Shaw, New York, NY: St. Martin's, 2000; Thomas Naff, *Conflict and Water Use in the Middle East. In Water in the Arab World: Perspectives and Prognoses* (ed.) Peter Rogers and Peter Lydon, Cambridge, MA: Harvard University Press, 1994.
5. For the full text, please see http://www.untreaty.un.org/ilc/texts/instruments/english/conventions/ 8_3_1997.pdf. The keyword 'equitable utilisation' as a principle gives no claim for an upper riparian state to have 'sole and exclusive' rights over the use of water derived from sources within their territory. For example, the Syrian and Iraqi (down-stream countries) claims for the use of water from the Tigris and Euphrates rivers that originate from up-stream Turkey are accepted under the 1997 UN Convention. Likewise, Syria's claim for absolute rights to the waters of the Jordan river without taking into consideration the downstream claims of Israel is unacceptable. The 1997 Convention does recognise the 'legitimacy of prior and historic user rights but the claim that historic use assures immutable and sole water rights is also not absolute in terms international law'.
6. Neda Zawahri and Sara Mitchell, 'Fragmented Governance of International Rivers: Negotiating Bilateral versus Multilateral Treaties,' *International Studies Quarterly,* 55, 2011, pp. 835–836.

7. Ibid., p. 836.
8. Uttam Kumar Sinha, 'Examining China's Hydrobehaviour: Peaceful or Assertive?' *Strategic Analysis*, 36(1), 2012, pp. 41–57.
9. Mariam Lowi, *Water and Power: The Politics of a Scarce Resource in the Jordan River Basin*. New York: Cambridge University Press, 1993.
10. Shlomi Dinar, Ariel Dinar and Pradeep Kurukulasuriya, 'Scarcity and Cooperation along International Rivers: An Empirical Assessment of Bilateral Treaties,' *International Studies Quarterly*, pp.55, 2011, 810, 812, and 814-815.
11. Ibid., p. 815.
12. Neda Zawahri and Sara Mitchell, 'Fragmented Governance of International Rivers: Negotiating Bilateral versus Multilateral Treaties,' *International Studies Quarterly*, 55, 2011, pp. 835-836; Marit Brochmann and Paul R. Hensel, 'The Effectiveness of Negotiations over International River Claims,' *International Studies Quarterly*, 55(3), 2011, pp. 859-882.
13. Shlomi Dinar, 'Scarcity and Cooperation along International Rivers,' *Global Environmental Politics*, 9, 2011, pp. 107, 109-110.
14. Robert Keohane and Joseph Nye, *Power and Interdependence*, New York: Pearson, 2011; Kenneth Pomeranz, 'The Great Himalayan Watershed: Agrarian Crisis, Mega-Dams, and the Environment,' *New Left Review*, 58, July-August, 2009.
15. Helga Haftendorn, 'Water and International Conflict,' *Third World Quarterly*, 21, 2000, p. 65.
16. Anders Jagerskog, 'The Jordan River Basin: Explaining Interstate Water Cooperation through Regime Theory,' *Occasional Paper No. 31*, Water Issues Study Group, SOAS, University of London, June 2001, at http://www.soas.ac.uk/water/publications/papers/file38374.pdf
17. *Atlas of International Freshwater Agreement*, UNEP, Nairobi, 2002, p. ix. This is a significant compilation of 'historical documents, statistical analyses and maps of the world's international basins and their agreements'. The compilation is a collaborative effort of the Oregon State University, the United Nations Environment Programme (UNEP), and the Food and Agricultural Organisation (FAO).
18. Of the 145 treaties, 124 are bilateral; 21 are multilateral, of which two are drafts. Such facts are found in the databank compiled by the Department of Geography and the Center for Freshwater Studies, Oregon State University in conjunction with the World Bank and the US Institute of Peace. The databank catalogues the treaties by basin, countries involved, date signed, treaty topic, allocations measure, conflict resolution mechanisms, and non-water linkages. For a detailed assessment and comparative analysis, see Jesse H. Hamner and Aaron T. Wolf, 'Pattern in International Water Resource Treaties: The Transboundary Freshwater Dispute Database', *Colorado Journal of International Environmental Law and Policy*, 1997 Yearbook.
19. Dinar, Dinar and Kurukulasuriya, n.10. p. 810.
20. Shlomi Dinar, 'Scarcity and Cooperation along International Rivers', *Global Environmental Politics* (9), 2009, p. 107.
21. Dinar, Dinar and Kurukulasuriya, n. 19, p. 812 and pp. 109 110.
22. This is based on the Oregon State University's river water treaty database. The analysis primarily looks at the river basins in Africa, West Asia, and Asia, at http://www.transboundarywaters. orst.edu
23. According to the 1967 Helsinki Rules, 'Each Basin State is entitled, within its territory,

to a reasonable and equitable share in the beneficial use of the waters of an international Drainage Basin'. The Helsinki Rules also defines the 'Drainage Basin' as 'a geographical area extending over two or more States determined by the watershed limits of the system of waters, including surface and underground waters flowing into a common terminus.'

24. Jaroslav Tir and John T. Ackerman, 'Politics of Formalised River Cooperation', *Journal of Peace Research*, vol. 46, no.5, September 2009, p. 623.
25. As noted by John Beddington, UK Chief Scientist, on 18, March 2009, at http://www.guardian.co.uk/science/2009/mar/18/perfect-storm-john-beddington-energy-food-climate
26. *Freshwater Under Threat: South Asia*, UNEP Report, 2008.
27. 'Population Growth and Per Capita water Availability in India', 1951, 1955, 1991, 2001, 2025 and 2050. Details available at: http://www.indiastat.com/table/percapitaavailability/24/watersupply/18198/365176/data.aspx. Also see, *Water: The India Story*, Report by Grail Research ; and *India's Water Economy: Bracing for a Turbulent Future*, World Bank Report, 2005.
28. Law, International Water. http://www.waterencyclopedia.com/La-Mi/Law-International-Water.html
29. The rule of 'equitable utilisation', is based on the concept that an international drainage basin is a coherent legal and managerial unit, and embodies a theory of restricted sovereignty under which each nation recognises the right of all riparian nations to use water from a common source and the obligation to manage their uses so as not to interfere unreasonably with like uses in other riparian nations.
30. While this is enshrined in the UN Convention on the Non-Navigational Uses of International Watercourses (1997), it is not binding because the Convention is not in force. Only 16 countries have ratified it. To come into force it requires 35 countries to have ratified it. The UNGA approved the UN Convention by 104-3 in 1997. There is also the 'no-harm rule' in the Convention, which states riparian nations to take all 'appropriate measures' to prevent causing harm to other watercourse nations.
31. Overall, in terms of per capita availability, China ranks among the world's lowest.
32. It is independent in terms of sources of rivers originating in its territory and not being dependent on the sources of the headwaters from other countries.
33. The only exception is one for basin cooperation on Lancang-Mekong River. The Lancang River begins in the northeastern side of Tanggula Mountains in Qinhai province, flows through Tibet into the Yunnan province, and is called Mekong River when it flows out of 244 boundary stone in Menglacounty in the Yunnan province. The Mekong runs all the way through Myanmar, Lao People's Democratic Republic, Thailand, Cambodia, and Vietnam. It empties into the South China Sea near Ho Chi Minh in Vietnam. The Lancang-Mekong River is 4,880.3 kilometers long; the section of Lancang River in China is 2,161.1 km long. So far, 25 major projects have been listed: 14 projects inside China, 4 inside Laos, 4 along the borders of Laos and Thailand and Laos and Cambodia, and 3 projects inside Cambodia. These dams are estimated to produce 70 per cent of China's current electricity needs.
34. 'Geochemistry of the suspended sediments of circum-Himalayan rivers and weathering budgets over the last 50 years'. See abstract http://adsabs.harvard.edu/abs/2003EAEJA....13617G
35. N.J. Ericksen, Q.K. Ahmad and A.R. Chowdhury, *Socio-Economic Implications of Climate Change for Bangladesh*, Dhaka: Bangladesh Unnayan Parishad, 1997.

36. J. Briscoe and U. Qamar, 'Pakistan's water economy running dry', *World Bank Report*, Oxford: Oxford University Press, 2008, p. 27. http://www.hec.gov.pk/InsideHEC/Divisions/FPD/cwf/Documents/Pakistan's%20Water%20 Economy%20Running%20 Dry%20Oxford%20University%20Press%202006.pdf
37. See, Jayshree Bajoria, 'The Question of Tibet', at http://www.cfr.org/publication/15965, December 2008.
38. The hydrological data of Tibet varies, and most of these figures are near estimates. See http://www.tew.org/tibet2000/t2.ch2.water.html
39. Ibid. Also see, An Caidan, *Travel Guide to Tibet of China*, Beijing: China Intercontinental, 2003, p. 13.
40. The National Development and Reform Commission had approved of the construction of 13 dams on the Nujiang River to generate 21,000 MW of power that would have been bigger than the Three Gorges Project. It was only because of Premier Win Jiabao's intervention (February 2004) that the massive projects were suspended. However, the possibility of restarting the project cannot be ruled out. See, http://news.xinhuanet.com/english/2009-05/22/content_11419960.htm
41. Reports indicate that Gezhouba, China's leading Construction Company, admitted to building a feedline for a dam on Yarlung at Zangmu. See, http://www.economist.com/blogs/banyan/2009/10/dammed_rivers.cfm. October 19, 2009.

2

HIMALAYAN HYDROLOGY: THE ANTHROPOCENE

We are in a geological epoch described as the Anthropocene. There is now no greater challenge to the well-being of the global commons than human-induced climate change. Since the industrial era began to trigger large-scale releases of fossil fuels, global average surface temperatures have risen by 0.8°C, already resulting in significant changes in physical, hydrological and ecological systems. Worldwide warming of 2-3°C above pre-industrial temperatures is very likely to herald major changes in terrestrial, marine and mountain ecosystems. These developments are all connected, and there is a risk of an irreversible cascade of changes leading us into a future that is profoundly different from anything we have faced before. We are gradually creating a hotter and less diverse world. 2014 may well be remembered as the year when climate change became understood as a current reality instead of a distant projection.

The Context

Security has a broader meaning than conflict – more so for an essential resource such as water. There are a range of assessments of the scope and scale of the problem of use, distribution, and scarcity of water in Asia, as often the perspectives adopted reflect political rather than scientific or legal considerations. A consensus is emerging that a broader understanding of demand side management is needed rather than a focus only on scarcity, seeking to 'securitise' water. This view questions approaches that look at resource scarcity solely in environmental terms, leading to the conclusion that managing changing water relations in Asia will be difficult. The need for trans-boundary cooperation will become even more important in order to build trust through mechanisms for sharing information, exchanging experiences, and using water and climate change as a basis for wider cooperation.

Profile of the Himalaya Mountain System

The Himalayas separates the Indian Subcontinent from the Tibetan plateau. The mountain system, as the Himalaya is referred to, includes the Karakoram, the Hindu Kush, and the Pamir Knot. As explained in the previous chapter, while geologically the Tibetan plateau is distinct from the Himalayan mountains, geologists often regard all the rivers, including those originating from Tibet, collectively as the 'circum-Himalayan rivers'.

The Himalayan glaciers, regarded as the 'Third Pole', contain one of the largest reservoirs of snow and ice outside the Polar region. Major Asian river systems – the Amu Darya, Indus, Ganga, Brahmaputra, Salween, Mekong, Yangtse, Yellow and Tarim – have their source in the Himalayan glaciers, contributing to almost 70 per cent of water resources. Almost 2 billion people – stretching from Afghanistan to the Ganga-Brahmaputra-Meghna (GBM) basin in South Asia to the Mekong Delta in Southeast Asia – are dependent on the flows of the rivers from the glaciers of the Himalaya that includes Tibet. The effects of global warming will be felt through the changes in the hydrological cycle. An effective adaptation policy cannot be delinked from the way water resources are managed and used.

In the 16th session of the Conference of the Parties (COP 16) in Cancun, November 2010, a new thrust towards understanding the water-climate link was visible. The Green Group of six countries – Cape Verde, Costa Rica, Iceland, Singapore, Slovenia, and UAE – encouraged greater emphasis on water management and climate adaptation. Unchecked climate variation can cause unprecedented challenges to the waterways of Afghanistan, Bangladesh, Bhutan, China, India, Myanmar, Nepal, Pakistan, and the countries in the Mekong basin.

The rivers originating from the Himalayan system will directly impact riparian relations. This can be analysed from three perspectives. First, the majority of the rivers are trans-boundary in nature, originating from, flowing through, and draining into territorially defined boundaries. Second, there is no clear definitional demarcation as to whether such a resource is exclusively a 'public good' (defined as non-rival and non-excludable) or a 'private good' (defined as rival and excludable)[1]. Third, planning any water resource utilisation policy will have to take into account the assessment of the impact of climate change in terms of seasonal flows and extreme events. River uses are deeply subjective in terms of where, what, and how they are being used. New modalities of water sharing agreements will have to factor in shifting hydrographs.

Himalayan Glacier Profile

Many studies using remotely sensed images have identified and analysed specific glacierised regions.[2] The mid-latitude, high-altitude Himalayan glaciers contain one of the largest reservoirs of snow and ice outside the Polar region, but with varied characteristics. Studies have confirmed that there is no 'stereotypical' Himalayan glacier.[3] While the glaciers in the Karakoram region of north-western Himalaya are mostly stagnating, those in the western, central, and eastern Himalayas are mostly retreating. Of these, the western glaciers have shown the highest rate of retreat.[4]

Scientific observations point out that the melting of 90 per cent of the Himalayan glaciers is directly caused by black carbon soot and other industrial processes. Other studies point to the presence of debris, such as pebbles and rocks, as an additional factor. The impact will gradually shrink glaciers, resulting in the decrease of water runoff in the long-term. In the short-term, earlier water runoff from glaciers when combined with seasonal rains can result in flood conditions. Many recent studies on the overall glaciers retreat and additional melt focus on water dammed or 'glacier lakes' that have the potential of generating dangerous outburst flooding.

In the coming years, rising concerns about the effects of climate change will take greater precedence over any other physical changes associated with climate change. First, as population grows, competition for food, energy, and water will correspondingly increase. Increasing demand for food grains will claim larger areas of cropland and greater volumes of irrigation water. Second, as noted earlier with the risks that climate change attaches, food-energy-water will be subject to various stresses and strains. Perceptions of a rapidly changing ecosystem may prompt nations to take several actions, many of them unilateral, to secure resources and territorial sovereignty. Any willingness to engage in greater river basin cooperation will depend on a number of factors, such as the behaviour of other competing countries, economic viability, and other interests that states are reluctant to either compromise or concede. Since disputes over water are inevitable because of the changes described above, understanding the processes of resolution as well as framing new mechanisms and approaches becomes a necessity.

Himalayan Hydrology

For almost half the world's population, water-related dreams and fears intersect in the Himalayas and on the Tibetan plateau. Other regions have their share of conflicting claims over water issues; but none combine the same scale of population, scarcity of rainfall, dependence on agriculture, scope for mega-dam projects, and vulnerability to climate change as those at stake within the greater Himalayan region. Here, glaciers and annual snowmelts feed rivers serving just under half of the world's population, while the unequalled heights from which their waters descend could provide vast amounts of hydro-power. At the same time, both India and China face the grim reality that their economic and social achievements since the late 1940s—both 'planned' and 'market-based'—have depended on unsustainable rates of groundwater extraction. Hundreds of millions of people now face devastating shortages.

It is being fast established that the Himalayan hydrology will be one of the critical frontlines in the global battle against climate change and water issues. The Himalayan mountain system is of crucial importance to the river system of Asia not only in terms of influencing the monsoon but also of the glaciers which are the source of many of the great rivers.

Of all the evidence showing the impact of global warming, perhaps none is more visible than, or as acutely dangerous, as outburst flooding in the Himalayas. According to the assessment of the Centre for Integrated Mountain Development (ICIMOD) in Kathmandu, there are about 200 glacial lakes in the Hindu Kush Himalaya region that are 'potentially dangerous': 25 in Bhutan; 77 in China; 30 in India; 20 in Nepal; and 52 in Pakistan. The ICIMOD keeps an inventory of 8,700 glacial lakes in the region. Glacial lakes are recognised as a threat to mountain areas worldwide. The lakes form as glacial melt-water collects behind ridges of loose rock debris called moraines that were deposited by the glaciers themselves.

In the last two decades, the impact of climate change on water resources cannot be discounted. In fact as a precautionary approach, the awareness of the dangers of climate change on water resources should frame future water policies in the region. Some of the studies/findings indicate increased precipitation in some areas, increased drought in some others, and increased variability of precipitation. Long-term trends for Himalayan glaciers under conditions of continued warming clearly point to melting, though some reports have tended to exaggerate the situation. The melting in the short term will help liberate melt-water which can be used for agriculture and industry. However, de-glaciation will also lead to rapid destabilisation of mountain slopes, causing landslides, rock-falls, and mudslides. This would directly impact the livelihood of the people who live on the floodplains of the major rivers spread across Nepal, India, Bhutan, and Bangladesh. The risks and uncertainties over the impact of climate change on water resources are potentially high in many South Asian countries

Climate Change and Water Resources

Though the amount of snow, ice, and permafrost on the Tibetan Plateau and its surrounding mountains, such as the Himalayas, Karakoram and Pamirs, is a lot smaller than that at the Poles, it is still significantly large. The glaciers covering the Himalayan region, including the Tibetan Plateau, is about 6 per cent of the area of the Greenland ice cap. Roughly, 1.7 sq km is permafrost, which is equivalent to 7 per cent of the Arctic's permafrost. The link between warming and glacial melting in the Himalayas is not clearly established. The Inter-Governmental Panel on Climate Change (IPCC) report in 2007, now well known, incorrectly suggested that Himalayan glaciers could disappear as early as 2035. In 2012, a study published in the international weekly journal of science, *Nature* showed that the glaciers in the Himalayas and Karakoram had lost little ice between 2003 and 2010, and that those on the Tibetan Plateau itself were growing.[5]

The study was further contested by scientists in the University of Zurich.[6] Further studies by the Third Pole Environment (TPE), with lead researchers Yao Tandong of the Institute of Tibetan Plateau Research, Beijing; Lonnie Thompson of the Ohio State University; and Volker Mosbrugger of the Senckenberg World of Biodiversity, Frankfurt, show that the area of the glacial lakes on the plateau has increased by about 26 per cent since the 1970s. The argument is about the correct measurement. The TPE suggests that using another satellite (ICESat) – which employs lasers to measure not only the areas of glaciers but also the elevations of their surfaces – one can come to the conclusion that, far from advancing, many of Tibet's glaciers are in retreat.

There are many factors that need to be investigated before drawing a conclusive link between warming and glacier melting. For example, over the decades, the Indian monsoon which brings snow to the southern part of the plateau and the eastern and central Himalayas has been getting weaker. The reason for this is yet to be concluded. Then, there is also wind strength and air temperature. Thus, the conclusion is the limited availability of data about the area. However, what is becoming increasingly clear is that, for South Asia, the monsoon, including the resulting snow fall in the upper reaches, is the determining factor in water flows from the Himalayas.

There is considerable attention on the Himalayan glaciers from the Western world. The US National Research Council has recognised the Hindu Kush Himalayas as a critical area. The Council's study draws upon scientific evidence to show that 'glaciers in the eastern and central regions of the Himalayas appear to be retreating at rates comparable to glaciers in other parts of the world, while glaciers in the western Himalayas are more stable and may even be increasing in size'. The study concludes that the consequences for the region's water supply are unclear. Their assessment is that shifts in the location, intensity, and variability of rain and snow due to climate change will likely have a greater impact on regional water supplies.[7]

Another assessment from the US Intelligence Community concludes that it is:

> prudent to expect that, over the course of a decade, some climate events—including single events, conjunctions of events occurring simultaneously or in sequence in particular locations, and events affecting globally integrated systems that provide for human well-being—will produce consequences that exceed the capacity of the affected societies or the global system to manage, and that have global security implications serious enough to compel international response.[8]

It assumes that the effective management of water resources will not take place with agriculture continuing to use approximately 70 per cent of the fresh water supply, thus 'posing a risk to global food markets and hobbling economic growth'.[9]

While the science remains uncertain, an effective adaptation policy for the South Asian states cannot be delinked from the way water resources are managed and utilised. The impact of global warming on water resources is particularly important for the Himalayan states that are highly dependent on the glacial sources of rivers in the Hindu Kush. Ongoing research indicates increased precipitation in some areas and increased variability of precipitation in others. Changes in precipitation and evapo-transpiration will greatly influence groundwater recharge. The expected decline in glaciers and snowfields will affect the flow of rivers, and increase the likelihood of floods due to overall increase in the intensity of rainy days.

A policy that awaits clear evidence may not be prudent. A precautionary approach and alertness to possible changes is a wiser option. The role of enforcement and monitoring agencies like the EIA (Environment Impact Assessment) needs to be effectively enforced in respective countries. The purposeful participation of civil society will be equally crucial for greater awareness and balance of development and water resources.

In the Himalayan region, the Glacial Lake Outburst Floods (GLOFs) problem, as explained earlier, is compounded by the fact that there is a lack of long-term data. Research on climate change impact on glacials need to be intensified at a regional level, and cooperation should entail sharing of data. The present state of knowledge is inadequate in identifying and assessing the magnitude of potential outbreaks of glacial lakes. GLOF risks have to be soberly assessed, and not heightened therefore leading to misperception. Countries in the region with trust deficit can easily misinterpret overstated risks, particularly downstream countries. Regional cooperation will need to factor in enhanced and updated forms of an automated early warning system. Also, upgraded remote sensing projects are important for flood warning systems because they can detect small changes in lake levels and send immediate signals to alarm systems near villages. Research and risk evaluation will also require ground-level surveys.

Another important feature which each individual state has to consider is to integrate and harmonise external water policies with internal water resource management.

The combination of rising population, increased urbanisation, and rapid economic growth compounds the challenges of securing water in the future. Though the most populous continent, Asia, has also the lowest water per capita in the world. Figures indicate that one in five people in the region do not have access to clean water. With an additional 500 million people expected in the next 10 years in the Himalayan watershed states, the stress on food, energy and water resources will only increase. It is, thus, important to understand the Himalayan region in terms of 'exponential function'—increasing population leading to greater food demand that increases dependence on water for irrigation and energy. The interconnection of food-energy-water is crucial, and if not framed sensibly into state polices, the cascading effect on food production, livelihood, and migration, will impact political stability in the region. What becomes worrisome is the likelihood of competition over water resources. Though the possibility of water being a direct cause of conflict is unlikely, however, given that the Himalayan watershed is fraught with tensions, water can act as a dangerous trigger and destabilise the region.

Himalayan Glaciers in South Asia

Himalayan glaciers constitute the largest body of snow, water, ice and permafrost outside the Polar region. Importantly, the glaciers provide fresh water to nine major river basins. Almost 1.5 billion people across India, Pakistan, Afghanistan, China, Nepal, Bangladesh, and several other countries depend on the glaciers and the monsoon for economic development and human needs. Increasing glacial melting in the Himalayas, as evidence suggest, is impacting water security in the region which, in turn, is causing concerns over human security and inter-state relations. The link between Himalayan hydrology, development, and regional security is becoming a challenge. While there remains considerable uncertainty about the extent to which overall glacier melting is occurring, in general, however, Himalayan glaciers are experiencing unprecedented rates of melting. Moreover, as compared to other mountains, glaciers are retreating faster.[10] Two prime reasons are cited: one the result of unprecedented global warming; and two, the increased levels of black carbon in the air.[11]

Some scientific evidence suggests the following:[12]

- A 43 per cent decrease is expected in glacial area on average by 2070, with a probability of 75 per cent decrease by the end of 21st century
- Average temperatures in the Indian Subcontinent are predicted to rise between 3.5°C and 5.5°C by 2100.
- Certain parts of the Himalayas are experiencing warming almost

eight times more than in the lower elevations. As compared to the global average, the warming is much higher.

- Increased warming is leading to more precipitation falling as rain instead of snow. In the higher elevations, precipitation is decreasing and hence a decline in snowfall. The rebuilding of glacier masses annually is slower.
- While warming and climate change is the key driver for the melting glaciers, other factors like black carbon are being identified as a significant contributor to the glacial melting. Some evidence suggests that black carbon may account for approximately 40 per cent of the melting in the Himalayas.

Moreover, increase in populations, demand for water use, inter-state tensions along with declining precipitation trends, droughts, and floods combine to put the region in an exceptionally precarious situation. The warning signs are now frequent challenging states in the region to cooperate and comprehensively deal with the issue that is complex and interlinked.

There has also been some unsubstantiated alarm-based analysis that the Himalayan glaciers might disappear by 2035.[13] Scientific investigations and further research has now focused on black carbon which is being increasingly found in the Himalayan snow as well as in the glaciers. Black carbon accelerates melting by contributing to atmospheric warming, and by increasing the amount of heat absorption within the snow by turning it black, allowing less to be reflected back out. Cement factories are the main sources of black carbon spread across the Himalayan belt. Adulterated fuels mixed with kerosene used by trucks as well as wood burning stoves also contribute to the black carbon. Evidence also suggests that some glaciers have actually 'grown' in surface area in recent years. Yet, there are further claims that surface area coverage often does not reflect total volume or changes in 'mass balance.'

The study on glacier melting is not complete as can be seen from the contradictory findings. Being a complex subject, glaciers defy generalisation. The fact that limited historical data on glaciers is available to use as a baseline suggests that further studies as well as research investment is required to understand changes vis-à-vis global warming and climate change. Some of the prominent glaciers for studies include the Kolahoi Glacier (Kashmir Himalaya), the Darang Drung Glacier (Ladakh Himalaya), the Chhota Shigri Glacier (Himachal Pradesh, central Himalaya), and the East Rathong

Glacier (Sikkim Himalaya). Glacier studies and monitoring; adaptation measures like locating and building water storages, opens up a new front for inter-state relations and regional cooperation.

Himalayan Glaciers and their Implications

With no uniformity in either melting or advancing, Himalayan glaciers are of unique importance to South Asia. The de-glaciation underway is already having downstream impacts. The first and most obvious concern with de-glaciation is the threat of excessive flow of water (floods) and diminishing water flows to the millions of people in the downstream regions of India, Pakistan, China, Afghanistan, Bangladesh, Nepal, and many other countries. Climate changes are also influencing changes in monsoon patterns, variation in overall precipitation levels, and the increased occurrences of extreme weather events.

Some of the facts below underline the criticality of glaciers and the monsoon in the Himalayan region.[14]

- Glacier melt contributes nearly 45 per cent of the water flows in the Indus River, and variable amounts of the water flow in the Ganga, Yangtze, Tarim, Mekong, Irrawaddy, and others. As some glaciers recede faster than others, certain river systems will feel the impact of diminished waters sooner than others.
- Rapid snow and glacier melting leads to floods, landslides, and rock slides, and also threatens the overall availability and temporal distribution of water. Glacial melting directly contributes to short-term problems of increased occurrences of glacial lake outburst floods (GLOFs) and long-term threats to water security for over 1 billion people.
- Monsoon precipitation is also projected to decrease in much of the region over the next several decades, directly impacting crop yields and food security; yet precipitation is projected to increase in other areas. In much of the monsoon regions, 50 per cent of the yearly total precipitation comes in just 15 days. Thus, any changes in these patterns can cause widespread problems for the populations that overwhelmingly depend on agriculture.
- Extreme weather events are already taking place more frequently, and are only projected to increase with the general warming trends.

China, India, Bangladesh, and Pakistan rank among the top 6 countries in the world in terms of vulnerability to floods.

- Food security is threatened not only by the direct loss of water for irrigating crops throughout the region, but also by a host of other factors as well – like direct damage from floods and other extreme weather conditions, etc.
- Water storage capacity in the region is extremely limited, resulting in extreme vulnerability of populations to fluctuations in water supply.

Inadequate Scientific Evidence

As observed from the evidence available, one can conclude that both glaciers and climate change are complex subjects. While the levels of uncertainty between actual and projected changes are wide, the fact remains that glaciological changes are impacting the Himalayan region with wide-ranging threats.

Thus, Himalayan glaciers have to be constantly monitored and studied in order to be prepared for potentially dangerous eventualities. Precautionary policies are better than waiting for accurate projections. Some of the areas that require further attention are changes in precipitation patterns, water flow fluctuation, extreme weather events, and the implications for food security. Some short-term potential benefits need to be observed and policies adapted. In fact de-glaciation, as observed, is resulting in the release of greater amounts of freshwater, particularly in the Indus River system. This near-term excessive flow will continue for three or four decades before declining rapidly. Adaptive measures – like building storage capacity and dams to protect against flooding and storing water for release during dry season – will be critical in the next decade. The selection of dam sites, the type of dams to be built, and the technology required will be critical. It will also be very crucial to get the inter-state politics and understanding right.

One of the major consequences of the glaciological-hydrological changes in the region will be the impact on food security. Loss or shortage of water for irrigating crops is the biggest threat to food security. However, there are also other factors, including crop losses, from rapid changes in temperature and precipitation, flooding, and other extreme weather events. It is estimated that 55 per cent of Asia's cereal production or 25 per cent of world production comes from the waters of the Himalayas and Central Asian mountains.[15] While the snow, water, ice and permafrost calculation will be a primary determinant, additional exacerbating factors in the region, including demography and the inter-state relations, will add to the overall assessment.

Glaciology and the Indus River System

The Indus River system is shared between India and Pakistan, and highlights a critical component of the relationship between the two countries. As mentioned, glacial melt contributes 45 per cent of water flow in the Indus alone. For Pakistan, it is a life line as 90 per cent of its freshwater is for irrigation. And, with 50 million people without access to safe drinking water, the water situation is grim. It is also estimated that with current consumption and management patterns, water availability in Pakistan is projected to decline 50 per cent to 70 per cent by 2050 and this is without factoring in any climate change or glacier melting.[16] Likewise, the situation in India is also challenging, and the scale is much higher. Water availability is expected to decline 40 per cent by 2050.[17] The additional exacerbating factors include: exponential increase in water demand from population increase, economic growth, and food and energy requirements; glacial and climate change impact leading to droughts and floods; low levels of water storage capacity and infrastructure, and high hydrological variability (within seasons and across seasons).

Adaptation Measures

Of the many river systems in the world, the mighty Indus with its tributaries is of striking and unmatched importance. The Indus basin has the largest irrigated area on any one river system. It comprises of the main river Indus and its major tributaries: the Kabul, Swat and Khurram from the west, and the Jhelum, Chenab, Ravi, Beas, and Sutlej from the east. There are three distinct physical features of the basin that needs to be noted: First, the Greater Himalayan ranges with their lofty peaks, snows, and glaciers, form a natural storehouse from which the rivers draw perennial supplies of water.

Thus, the impact of climate change on the glaciers is critical to the future flows in the rivers. Second, the physiography of the Lesser Himalayas and the Shivalik as sources for the development of hydroelectricity is potentially high. This is because the rivers of the Indus system receive all their waters only in the upper parts of their mountainous catchments, and have maximum flow when emerging from the foothills. Third, the basin also covers Tibet from where the Indus and Sutlej originate as well as Afghanistan from where the Kabul begins. China is the ultra-upper riparian, and given the capacity and capability it has to harness or divert the waters of the Indus and Sutlej, it can change the hydrological dynamics of the basin. Comparatively, Afghanistan is less significant as an upper riparian, given its political instability.

A study of the hydro-electric schemes by the Central Electricity Authority of India[18] suggests that the Indus basin has a potential capacity of 19988 MW at a 60 per cent load factor, which is the highest in India. If the projected share of the hydroelectricity is to improve from 25 per cent to about 32 per cent by 2025-26,[19] the Indus basin would be a prime target area. In Pakistan, hydel-power is the second largest source of electricity, contributing 33.1 per cent of total power generation. While Pakistan's hydel generation potential is estimated to be 46,000 MW, only 14 per cent has been exploited.[20] Unlike India, Pakistan is a one-river-basin country, and its entire hydel-power projects falls under the Indus.

Another issue that has received extra attention was that of the extremely limited water storage capacity in the region. For example, India and Pakistan store less 250 cubic meters of water per capita compared to more than 5,000 cubic meters per capita in the USA and Australia. The lack of water storage capacities leaves the already vulnerable populations at great risk to fluctuations in water flows and changes in monsoon patterns.

Storage capacity is one area where states in South Asia can together take some concrete steps to respond to water security challenges. Investments can be ramped up to increase both the natural and constructed water storage systems. Natural storage systems include snow, ice, glacial lakes, wetlands, and groundwater aquifers. Constructed systems include artificial ponds and tanks, large reservoirs (behind dams), and human modifications to the natural systems including large-scale watershed management programmes, groundwater recharge systems, Karez irrigation systems, and temporary runoff collection areas. A strategy for climate change adaptation in the Himalayas is an endeavour that South Asia cannot miss.

Some other adaptation measures beyond the water storage capacity could include the following:[21]

- Efficient irrigation systems in the plains
- Building institutional capacity to manage increasing variability of weather and water flows
- Implementing early warning and preparedness systems for extreme weather events
- Developing protective infrastructure and selecting sites for embankments, cyclone shelters, check dams, etc.
- Using cellular communications to warn of flooding
- Policy emphasis on agricultural adaptation
- Improving water use efficiency and productivity of irrigated agriculture
- Assisting with income diversification to reduce vulnerabilities
- Supporting regional/intra-state river basin institutions and/or regulatory authorities

Research and Monitoring

The region would require scientific assistance to assess and monitor the physical changes in the region. Ideally, the region should come together to set up its own satellite and tracking systems and, if required, seek initial help through National Aeronautics and Space Administration (NASA) and Geographic Information System (GIS) as well as satellite imagery and related scientific cooperation programmes. Equally important will be scientific and technical investment in the development of early warning and preparedness systems, like the Hydro-met. Lessons learned in such exercises across the mountains – for example, in the Andes or the Alps – should be incorporated to confront the projected melting trends in the Himalayas.[22]

Any impetus has to come from the region. India's leadership and its recent policy initiatives in the region are welcome. Outside role could be those the World Bank or ADB as facilitators or technical brokers in sharing information and experiences from other cases of conflict and cooperation over water resources management. The world is a river, and each river basin has valuable lessons for other basins.

Himalayan Hydro-politics

Given the distances that the Himalayan rivers traverse in their respective territories, China and India are critical players in the hydro-politics of the region. India is simultaneously an upper, middle, and lower riparian. China's hydrological position, on the other hand, is one of complete upper riparian supremacy. India's middle riparian position, increases its dependency (and thus water insecure) on the headwaters of the rivers such as Indus, Sutlej and Brahmaputra which originate in the Tibetan plateau. China is equally water insecure, but its insecurity relates to the disproportionate availability or uneven distribution of waters within its territory, the majority of which is in the south (Tibet Autonomous Region), with the north and west excessively water stressed. In terms of per capita water availability, China ranks amongst the world lowest.

Water pollution is also a big worry, with the vast majority of lakes and many of its largest rivers unsuitable for human use. The territorial source of the Himalayan rivers makes China far more water secure than India. It would be probably fair to say that China is the world's most independent riparian country.[23] This hydrological position gives it enormous latitude in shaping larger political equations with its riparian neighbours. On the other hand, given its middle riparian position and its longstanding commitment to bilateral river treaties, India has to assiduously balance the anxiety and concerns of its lower riparian (Pakistan and Bangladesh) without compromising its own water requirements.

In contrast, China has no bilateral riparian treaties,[24] and is, therefore, not bound by any water utilisation agreement with its neighbours. In fact, China was one of the three countries[25] that did not approve of the 1997 UN Convention on the Law of the Non-Navigational Uses of International Waterways. On the Mekong River basin, China is only a dialogue partner in the Mekong River Commission (MRC), which was formed by the concerns of four lower-riparian countries – Cambodia, Laos, Vietnam and Thailand – in 1995. Though China's non-binding participation in the Mekong basin has increased, it is unthinkable that it will join MRC as an active member.

For the Himalayan basin-states other than China, being water-dependent from external sources is a hydrological reality. The water dependency and the prevailing politics shape fears and perceptions. For example, the Mekong lower riparian countries will remain suspicious of China's upstream hydroelectricity projects.[26] In the case of Pakistan (which is heavily dependent on the sources of rivers outside its boundary), India (and not China) is seen as an upper riparian aggressor. This, of course, greatly relates to the grievances that Pakistan has over the Indus Water Treaty with India. Similarly with Bangladesh, the lowest riparian in the GBM basin, water becomes a political and emotional driver. As the upper riparian player, China would like the water debate in Pakistan and Bangladesh to be directed and contested with India, without highlighting its own hydroelectricity plans, either on the Indus or the Yarlung Tsangpo (Brahmaputra in India).

While China has no formal water sharing arrangements with its neighbouring countries, India has several treaties to address water issues with its neighbours – like the 1960 Indus Water Treaty with Pakistan, and the 1996 Ganga Treaty with Bangladesh. With Nepal and Bhutan, treaties have been signed to share the benefits of water. Water treaties commit India to a dialogue-based water sharing approach, and become an important part of its nighbourhood policy diplomatically. In contrast, China would tend to take a strategic view of the water it commands and, given its hydrological position, factor water as a tool, leverage, and a bargaining instrument in framing its regional policies. A snapshot of the riparian dynamics in the Himalayan watershed suggest that while there is considerable lack of trust on water issues between states, there is greater possibility of drawing India – rather than China – into the regional water debate and break political deadlocks through sensible water sharing arrangements and resource development.

NOTES

1. Jaroslav Tir and John T. Ackerman, 'Politics of Formalised River Cooperation', *Journal of Peace Research*, vol. 46, no. 5, September 2009, p.623. There is, however, a general view to perceive water resources as 'collective goods' or 'common pool resources'.
2. http://www.treehugger.com/files/2010/08/usgs-confirms-himalayan-glaciers-melting climate-change-to-blame.php#
3. Study carried out by the Department of Geography, UC Santa Barbara. See *Hindustan Times*, 27 January 2011. Also see, http://www.sciencedaily.com/releases/2011/01/110124162708.htm
4. Ibid.
5. Thomas Jacob, John Wahr, W. Tad Pfeffer and Sean Swanson , 'Recent Contribution of Glaciers and Ice Caps to Sea Level Rise', *Nature*, 482, February 23, 2012, pp. 514-518
6. Tobias Bolchfrom the University of Zurich argued that Jacob's finding was based on seven years of measurements by a satellite mission called the Gravity Recovery and Climate Experiment (GRACE). This used orbiting gravimeters to try to measure changes in the

ice cover from effects on the local gravitational field. According to Bolch, the approach suffered from two problems. One was the coarse resolving power of the satellite's instruments. These could not detect changes in features less than 200 km across. This is enough to study large regions with homogenous surfaces, such as the Arctic and the Antarctic (which GRACE did, in fact, do). But mountainous terrain has complex topography. The second, more serious problem is that the satellites cannot tell the difference between solid and liquid water. If a glacier melts but the water stays put as a lake, GRACE will see no change. Since the Tibetan Plateau contains a lot of 'closed' catchments from which melting water cannot easily escape, large amounts of melting could happen without GRACE detecting them. See, http://phys.org/news/2014-01-meltwater-tibetan-glaciers-pastures.html

7. 'Himalayan Glaciers: Climate Change, Water Resources and Water Scarcity', Report by the National Research Council, USA, 2012, http://dels.nas.edu/Report/Himalayan-Glaciers-Climate-Change-Water-Resources/13449
8. 'Global Water Scarcity', USA Intelligence Community Assessment, ICA 2012-08, 2 February 2012. See, http://www.transboundarywaters.orst.edu/publications/publications/ICA_Global%20Water%20Security[1]%20(1).pdf
9. Ibid.
10. See, Waters of the Third Pole: Sources of Threat, Sources of Survival, at https://www.chinadialogue.net/UserFiles/File/third_pole_full_report.pdf. Jane Qiu, 'China The Third Pole', at http://www.nature.com/news/2008/080723/full/454393a.html
11. See, Waters of the Third Pole: Sources of Threat, Sources of Survival. https://www.chinadialogue.net/UserFiles/File/third_pole_full_report.pdf
12. There are many reports and assessments on the Himalayan glaciers and the likely impact of climate change 'Himalayan Glaciers, Climate Change and Water Security in South Asia', Synthesis Report, Bureau of Central and South Asian Affairs, U.S. Department of State, 17 November 2009.
13. 'IPCC officials admit mistake over melting Himalayan glaciers', *The Guardian*, 20 January 2010, at. http://www.theguardian.com/environment/2010/jan/20/ipcc-himalayan-glaciers-mistake
14. 'Himalayan Glaciers, Climate Change and Water Security in South Asia', Synthesis Report. Bureau of Central and South Asian Affairs, US Department of State, 17 November 2009. Also see, 'Climate Change 2013: The Physical Science Basis', IPCC WG1https://www.ipcc.ch/pdf/assessment-report/ar5/wg1/WG1AR5_SummaryVolume_FINAL.pdf
15. Centre for Integrated Mountain Development (ICIMOD), 2009.
16. 'Climate Change Impacts and Vulnerability in the Eastern Himalayan', ICIMOD, 2009, http://lib.icimod.org/record/26800/files/attachment_675.pdf
17. Ibid.
18. See http://www.cea.nic.in/hydro/Special_reports/Ranking_Study/preliminary_ranking_study of hyd.htm
19. Ibid.
20. *Economic Survey of Pakistan 2007-2008*, at http://www.finance.gov.pk/admin/images/survey/chapters/15-Energy 08.pdf, p. 248.
21. 'Himalayan Glaciers, Climate Change and Water Security in South Asia', Synthesis Report, Bureau of Central and South Asian Affairs, US Department of State, 17 November 2009. Also see, 'Himalayan Glaciers: Climate Change, Water Resources, and Water Security',

National Research Council of the National Academies, Washington DC, 2012, at http://dels.nas.edu/Report/Himalayan-Glaciers-Climate-Change-Water-Resources/13449

22. 'Mountains and Climate Change: From Understanding to Action', Centre for Development and Environment (CDE), Institute of Geography, University of Bern, 2012, at http://www.fao.org/docrep/017/i2869e/i2869e00.pdf
23. 'Independent' in terms of rivers originating in its territory, and not being 'dependent' on the headwaters from other countries.
24. This is except for being a dialogue partner in the Mekong River Commission (MRC). The Lancang River begins in the northeastern side of Tanggula mountains in the Qinhai province, flows through Tibet into the Yunnan province, and is called Mekong River when it flows out in Mengla county in the Yunnan province. The Mekong runs all the way through Myanmar, Lao People's Democratic Republic, Thailand, Cambodia, and Vietnam; it empties into the South China Sea near Ho Chi Minh in Vietnam. The Lancang-Mekong River is 4,880.3 kilometers long; the section of Lancang River in China is 2,161.1 km long. So far, 25 major projects have been listed: 14 projects inside China, 4 inside Laos, 4 along the borders of Laos and Thailand and Laos and Cambodia and 3 projects inside Cambodia. These dams are estimated to produce 70 per cent of China's current electricity needs.
25. The other two were Turkey and Burundi.
26. Likewise, Kazakhstan and Russia will be concerned over China's diversion of the Irtysh and Ili rivers.

3

Ganga Basin and Regional Cooperation

The Ganga

Possibly one of the most modernising policy approaches of the current BJP-led national government has been to raise the profile of rivers. Water is indispensable to governance and development plans, as well as livelihood and a healthy life. These have been expressed as *sujalam sufalam* (water for prosperity) and *swachh Bharat abhiyan.* Rivers are also a key instrument of regional prosperity and integration. The earlier Ministry of Water Resources has now been expanded to include River Development and Ganga Rejuvenation. Likewise, Drinking Water and Sanitation, which was a department under the Ministry of Rural Development, is now a full-fledged ministry with a cabinet rank minister. These are early indicators, hopefully, to a more serious rethinking, readjustment, and enhancement of approaches and technologies regarding water, especially since various water projects are witnessing widespread negative ecological and livelihood impacts.

A stable supply of quality water is an essential requirement for guaranteeing socio-economic-political stability. The Ganga has become central to river conceptualisation, cleaning and planning in India. With a length of 2,525-kms, the Ganga is the longest river of India, and the second largest river in the world in the amount of water discharge. It also carries one of the highest sediment loads.[1] The Ganga basin occupies a quarter of India's land mass, and is one of the most populous regions in the world. However, with a faecal coliform[2] countdrastically overshooting the quality limits in some stretches, the water is well below the health limits for drinking, bathing, and even irrigation. The Ganga is the 5th most polluted river in the world. While this remains a shameful reality, it is important to continuously reinstate the symbolism of the Ganga as the river of faith, devotion and worship. Rabindra Nath Tagore made several references to the Ganga in his works, particularly *Jiban Smriti* and *Rabindra Rachnabal*, by expressing the Ganga as the sacred thread of Indian *yajna* – connecting years of wisdom, religion, and meditation.[3] There is a Ganga in every part of India, and the emblematic interpretations of the river can become a collective force of change.

The Ganga and many other rivers present numerous difficulties in the way of effective development and management. Moreover, the cleaning of the rivers cannot be separated from rejuvenating their flow. But rivers also need to be harnessed for the wider national needs. Economic development and the requirements of energy and food cannot be divorced from developing rivers in terms of storage for irrigation and electricity generation. This is critical as India's urban population (419 million) will double by 2050.[4] The 'smart cities' plan in 100 cities across India, cannot be sustainable without more intelligent human-nature symbiosis. The challenge is to balance environmental, human, cultural, pilgrimage, and economic interests within the complex system of a major river.

Any plans regarding a river have to see it as a whole, and not just in terms of its parts. Perhaps the most crucial is basin planning, with attention being given to marginalised rivers (tributaries and rivulets) and not just to the mainstream. The upstream, middle course, and the delta of any river system will require greater efforts in terms of integrating and harmonising with careful study, data analysis, as well as federal cooperation and consultation. One of the neglected aspects of river development in India has been navigation. The golden age of navigation in the Ganga was mid-19th century, with steamers of the East India Company plying as far upstream as Garhmukteshwar. However, with the development of the upper Ganga canals and the railways, navigation was effectively grounded. This needs to be revived. Using connected rivers for logistics and freight movement is a cleaner form of transport. Importantly, a thrust towards navigation will also drive planning towards maintaining flow as well as a relook at the development of the upper catchment area.

Nearly one billion people[5] (one in seven of the world's population) live in the three river basins of the Brahmaputra, Ganga and Indus. These rivers are central to the lives, livelihoods and food security in the region and to the maintenance of the natural environment. Balancing the needs of these different users is an enormous challenge particularly as populations grow and climate change increasingly alters the predictability of river flows and access to water. Management of these water resources is by nature transboundary and historical distrust and prioritisation of national interests have prevented the emergence of coherent and effective transboundary management systems able to balance the needs of those who rely on these resources the most. Importantly, the men and women most reliant on these water resources have the least access to information on how these resources are managed, who is responsible and what they can contribute.

This chapter examine India's hydrological experience with Nepal and Bangladesh in sharing the waters and the benefits of the Ganga. Positive interventions on the Ganga will not only provide opportunities for closer regional integration but also provide an outcome to help resolve political issues. The Ganga, thus, becomes a catalyst for transforming bilateral friction to tangible gains.[6]

Riverine Collaboration

The criss-crossing of rivers in South Asia makes it a predominantly riverine region in which a regional approach and cooperation on the rivers is critical to well-being. For example, the cleaning Ganga *abhiyaan* cannot be just accomplished with India's participation. Nepal too has a significant role to play, given that many tributaries of the Ganga originate in the upper riparian state, especially in the dry season. A sustained flow or the ecological flow (e-flow) is critical to the health of the river. A reduced flow leads to a significant drop in the dissolved oxygen (DO) level, thus dangerously affecting the self-purifying ability of the Ganga. It is observed that the current level of pollution waste into the Ganga is beyond the river's capacity to self-purify.[7] Various tests and measurements confirm that the DO level in the Ganga has dropped to 4 mg/l (milligram per litre).[8] This is alarming as the threshold level is 5 mg/1. Importantly the water supply to the Ganga depends on the monsoon. The river receives 75-80 per cent of total annual rainfall in a three month period from mid-June to mid-September. Thus, the monsoon becomes a vital source of river flow. Equally, the melt of the glaciers becomes crucial in the dry season. Augmenting the river flow has always been a hydrological challenge and a source of constant contention between India and Bangladesh. Improving the river flow to benefit Bangladesh would also mean the active participation of Nepal in the Ganga basin. Bangladesh, the lower riparian sharing 54 rivers with India, is much stressed by the reduced river flow as well as pollution. This has not only led to mistrust and acrimony between the two states, but also created potential security concerns for India, including border-related issues, cross-border migration and water sharing disagreements.

The challenges and prospects of water management issues of the Ganga Basin[9] remain a core component of India's bilateral relations with Nepal and Bangladesh. Nepal's hydro-relations with India date back to the Exchange of Letters of 1920 with the then British Government on the construction and operation of the Sarada Barrage Project.[10] This is regarded as the first bilateral agreement in the Indian Subcontinent.[11] Through a give-and-take format, Nepal agreed to transfer 4093.88 acres of her land on the east bank of the Mahakali to India so that the latter could build the Sarada Barrage for irrigation. In exchange for the land, Nepal received an equal amount of land elsewhere. In addition, India agreed to provide Nepal with a 'supply of 460 cusecs of water and, provided the surplus was available, a supply of up to 1000 cusecs from the Sarada canal should cultivation expand at any time in the future'.[12]

Post-independence, India signed the Peace and Friendship Treaty with Nepal in 1950, which established the framework for a unique relationship on water resources. As part of the framework, India cooperated with Nepal on the Kosi (1954) and the Gandak (1959) dam projects to control flooding and help irrigation in Bihar and Uttar Pradesh.[13] The Kosi project was undertaken to mitigate the recurrent and devastating effect of the Kosi flood during the monsoon season. The Kosi multipurpose scheme consisted of a canal system, flowing channels on both sides, a barrage across the river, and a hydro station. However, the project was seriously criticised at all levels in Nepal: the prevailing feeling was that it was a sell-out of national property for India's benefit, with no gains for Nepal. What was considered as mutually beneficial hydro-cooperation soon became the cause of bitter resentment within Nepal, and which even today greatly shapes the perception of India-Nepal relations. Based on some of the concerns of Nepal, a revised agreement was signed in 1966.

The Gandak Project Agreement (1959) also emphasised flood control, as well as irrigation, with hydropower as a secondary benefit. The fate of this agreement was also similar to that of the Kosi agreement. It was severely criticised by Nepali politicians. Like the Kosi Agreement, this one too was revised, with some additional benefits given to Nepal. Subsequently, both parties agreed to revise this agreement and, after its revision, some more benefits were given to Nepal. Nonetheless, in relation to the criticism, it has also been argued that in both the Kosi and Gandak projects India had invested a huge amount of money, manpower, and technology and, therefore, it was only natural that more benefits should accrue to it. In other words, the hydro-relation which was projected as a sharing of benefits and thus help build a stable political relationship between the two countries ended up creating a dysfunctional relationship between them.[14]

Likewise, India's relations with Bangladesh have been cramped by water management issues since Bangladesh became independent in 1971. The search for the long-term augmentation of the water flows of the Ganga is an important context in the riparian relations of India and Bangladesh.[15] The World Bank's South Asia Water Initiative (SAWI) describes the Ganga hydrologically as a 'complex interplay of run-off, glacier and snow melt, and ground water acquifers compounded by pronounced seasonality and climate variability'.[16] The annual precipitation in the Ganga Basin is high for three months a year (mid-June to mid-September), while in the rest of the year, the area remains water-stressed. Improving the river flow or augmentation is, thus, critical for bilateral relations. Reduced river flow and pollution has affected Bangladesh, which has not only led to misunderstandings and tensions between the two countries, but has also generated existing and potential security concerns for India, including border disputes, water disagreements, and illegal cross-border migration.

In spite the establishment of the Joint Rivers Commission in 1972,[17] water issues have remained divisive and emotive. Bangladesh has long accused India of drawing too much water from the Ganga, which has led to negative impact on irrigation, navigation, water quality, and ecology in southwest Bangladesh. In 1996, the political environment was conducive to concluding the Ganga Agreement.

With its regional thrust and accommodative approach, the Janata Dal-led government in India was receptive to water resource development and water sharing agreements. But more importantly, New Delhi had a found in the Awami League (AL) a compatible political party on which it could build future relations. For the newly-elected AL in Bangladesh, nothing could have been more significant than the signing of the Ganga Treaty.[18] However, the critical issue of the augmentation of flow during the dry season still remains unresolved. The challenge is how to augment the flow? This is critical – more so with the increased withdrawal of water and overall increased consumption. The issue of climate change and global warming are equally impacting the flow. Thinking about augmentation would mean taking a basin approach which makes the issue not merely a bilateral one between India and Bangladesh but also brings in Nepal and Bhutan. This means that the low flow in Bangladesh can be turned around by creating water storages in Nepal and Bhutan that can be released in the dry season. The thinking process has to go beyond the bilateral to 'from the source to the mouth'. It will require proper site selection and assessment; right technological intervention; and societal participation.

Ganga in Nepal-India Relations

The Ganga originates in India; but a significant portion of the volume of its waters is contributed by flows originating from Tibet (through Nepal) or from Nepal itself. According to the Water and Energy Commission Secretariat (WECS) of Nepal, there are 6000 rivers in Nepal with a total drainage area of 194,471 sq. kms and an estimated annual run-off of 220 billion cubic metres.[19] Rivers in Nepal are classified by their origins:

> The first category comprising the four main river systems of the country – Koshi, Gandaki, Karnali and Mahakali – originate from glaciers and snow-fed lakes. Rivers of the second category – Babai, West Rapti, Bagmati, Kamala, Kankai and Mechi – originate from Mahabharat range. Streams and rivulets originating mostly from the Chure hills make up the third category; these rivers cause flash floods during monsoon rains, and remain without any flow or very little flow during the dry season.[20]

The surface water availability in Nepal is about 225 billion cubic metres (BCM) per annum, of which only 15 BCM per annum is in use. The Ganga is the natural drainage of all the rivers flowing in from Nepal, and various sources suggest that the Nepal rivers contribute 46 per cent of the flow of the Ganga. A critically important fact is that, during the lean season from March to May, the contribution of the flow increases to about 75 per cent.[21] Another important fact is that of the 1,880 kms long India-Nepal border, only 640 km is river boundary. Given its mountainous terrain and upstream position, Nepal is hydrologically crucial not only for hydropower development but also – and more importantly – for downstream flood control and dry season augmentation.

Nepal has abundant water resources; it also has favourable sites to build large dams that are capable of storing up to 77 billion cubic metres of water. Estimates indicate that this storage size is about 68 per cent of total monsoon flow. Thus, it becomes essential that Nepal cooperates with its lower riparian neighbours, India and Bangladesh, to meet their water demands during the dry season. In this context, the willingness of Nepal's King Birendra Bir Bikram Shah Devto utilise Nepal's vast water resources for the region is interesting. While addressing a gathering in Kathmandu in 1977, the late King referred to the vast water resource of Nepal as a 'collective benefit of all the people of the region'.[22] It was probably for the first time that a regional approach to water was initiated. However, given the politics of the time, India approached the issue from a bilateral perspective. As the lower riparian, Bangladesh, quite naturally, welcomed the idea. Again, at the South Asian Association for Regional Cooperation (SAARC) held in Dhaka in 1985, King Birendra emphasised the need for regional cooperation in the field of water resources. He said:

> From the side of Nepal with the high Himalayas as one of our assets and a vast reservoir of yet untapped water resource that can give to the millions of our people a means to fulfil their basic needs, I wish to draw the attention to the fact that there exists this priceless resource waiting to be harnessed for the benefits of our people.[23]

Bilateralism has remained the bedrock of India's neighbourhood policy. A few months before the 1st SAARC Summit in Dhaka in 1985, India's Prime Minister Rajiv Gandhi stressed on some central positions that India would not be willing to negotiate. He said, "We have not sought to melt our bilateral relationship into a common regional identity, but rather to fit South Asian cooperation in our respective foreign policies as an additional dimension".[24] On India's insistence, all political or security matters were excluded from the Summit.[25] However, the Dhaka Declaration described the meeting as a 'tangible manifestation of their determination to cooperate regionally'.[26] China and Pakistan are factors in India's reluctance in engaging in regional cooperation. Another factor is that India wants to maintain flexibility in its negotiations with its neighbours. Bilateral benefits from hydro-relations are, thus, acceptable to India, and therefore India has set up Joint Committees and Commissions with Bangladesh and Nepal separately. In other words, India has different strategies and approaches, a mix of lower riparian and upper riparian positions, in dealing with Nepal and Bangladesh on the water issues.

The history of ideas and initiatives for regional cooperation in South Asia has always been problematic. Clearly, as the largest basin country, India's approach will play a vital role towards regionalism. So far, India has been pursuing a strategy of bilateralism with her neighbouring countries in the water sector. However, the current BJP-led government neighbourhood emphasis based on riverine thinking and comprehensive basin management demonstrates a shift in the policy mindset, and a movement towards a sub-regional framework.

Hydropower Cooperation

Nepal is the second richest country in inland water resources; but its huge hydropower potential remains grossly unutilised. Consequently, it has led to chronic power shortages and stymied economic development. About 60 per cent of the population have no access to electricity. The country remains one of the world's poorest, with firewood accounting for three-quarters of the energy consumed. Nepal's hydropower potential is estimated to be 83,000 MW, of which 40,000 MW is financially and technically feasible. However, it has only been able to exploit 680 MW.[27] Nepal's limited capacity to harness its water resources pushes it to cooperate with India which has advanced hydropower capabilities. India's huge energy requirement and its close proximity to Nepal make it an ideal country for cooperation on hydro development.

As mentioned, India and Nepal hydro-cooperation began in 1920 with the Sarada Barrage project on Mahakali river, which later Nepal felt was disadvantageous. In fact, many in Nepal view the Sugauli Treaty of 1816, imposed by the British East India Company, as being unequal, and demanded the restoration of the illegally occupied territory.[28] The historic hurt framed as 'give-away' of the 1816 Treaty continues to haunt India-Nepal hydro-relations.

In the 1950's, India signed with Nepal two major water projects, the Kosi Agreement (1954) and the Gandak Irrigation and Power Project (1959). However soon, India's strategic and security considerations as well as its hydro-cooperation with Nepal led to deep-seated misperception and mistrust. Clearly, any water cooperation with India is eventually seen as a 'sell-out' in Nepal and often raises the following questions: What is it about the Nepalese mindset that perpetually remains chary of India thus creating a dysfunctional relationship? Is it more to do with domestic political squabbling and the lack of consensus amongst the Nepalese political parties? Is this to draw China into its water development? Is this the result of a small landlocked country that has two big and powerful countries as its neighbours (China and India)? These are some of the interesting and perplexing questions that the relationship between India and Nepal has to grapple with.

In both the Kosi and Gandak Agreements, the Nepalese felt that India had secured disproportionate benefits at Nepal's cost. An example often cited by Nepal is the dam on the Kosi river, built by India after signing the Kosi Project Agreement. The dam flooded a large piece of territory in Nepal, but helped India irrigate most of the agricultural land in Bihar. Nepal frequently complains that India's construction of the dam was an encroachment on Nepal's territory.[29] Similarly, on the Gandak Project, Nepal claimed that India had taken advantage of the agreement by diverting the flow of the Gandak river to serve its own vested interests. Both the 1954 and 1959 water agreements have become an emotive issue in the bilateral relations between the two countries, with different interpretations regarding the benefits accrued.

Hurdles to Cooperation

The historical experience of water resources cooperation between India and Nepal is vexing, and greatly impedes future plans. Even the two treaties signed in the 1990s – the Tanakpur Agreement (1991) and Mahakali Treaty (1996) – have suffered similar mistrust, with questions of 'fairness' being constantly raised by Nepal. The Mahakali Treaty, signed by then Nepalese Prime Minister Sher Bahadur Deuba and then Indian Prime Minister P.V.Narasimha Rao, was planned for a multi-purpose joint development of the Mahakali River on irrigation, flood control, and power generation. The treaty specified that the Pancheshwar Multi-Purpose Project would produce 6,000 MW of hydropower by 2002. Nepal soon showed resentment, and like in the past, came to consider the treaty as being lopsided. Further, the growing civil society constituency have cautioned that large-scale hydropower projects are detrimental to the ecosystem, leading to severe landslides and flash floods.

Despite historic difficulties, hydropower development remains a driver in India and Nepal relations, particularly as Nepal moves towards political stability as well as India's emphasis on the neighbourhood. A string of events complements this: in August 2013, Kathmandu decided to re-commence the 900 MW Upper Karnali project, and in 2014 the GMR Group signed the Project Development Agreement with Nepal.[30] Another Indian company, Coastal Projects Limited, has been given permission by Nepal to invest in the 44MW Super Madi Hydropower project in Kaski district.[31] At the SAARC Summit in Kathmandu in November 2014, India and Nepal signed yet another agreement for the 900MW Arun III dam, making India the biggest hydropower developer in Nepal.[32] India's emphasis in the region has been well noticed and since 2013 efforts to multilaterally cooperate as NIB (Nepal, India and Bangladesh) not only on hydropower but also jointly financing projects in the Ganga river basin have gained traction.[33]

Cleaning the Ganga

In its important and well recognised study, the Centre for Science and Environment (CSE)[34] put forward two main policy approaches to cleaning the Ganga.[35] According to the study, a 'pre-emptive approach' is to stop domestic sewage and sullage and industrial effluent from reaching the Ganga. A 'pro-active approach', as the study mentions, is to maintain a minimum level of flow within the Ganga, especially during the dry season, which will allow the river to continuously self-purify itself. An increased flow or augmentation will also encourage navigation.

The 'pre-emptive approach' allows for India to collaborate with Nepal on checking pollution in the Ganga, particularly towards establishing sewage treatment plants (STPs) along the four main upstream rivers – the Mahakali, Karnali, Gandak and Kosi. Such upstream intervention has been similarly undertaken by India under the Ganga Action Plan (GAP) I and II.[36] Efforts towards cleaning the Ganga demonstrate that trans-boundary rivers have an ecosystem wholeness that cannot be segmentised.

In the 'pro-active approach', India and Nepal have equally significant scope for collaboration on storage facilities in the highlands of Nepal during the monsoon period from June to November. The excess water stored can then be released to augment the low flow in the Ganga during the dry period between December and May. Storage facilities can have widespread benefits. These include cleaning the Ganga east of Patna, generating energy, and managing the flood problem in the state of Bihar and Uttar Pradesh during the monsoon seasons. The release of stored water in Nepal during the dry season will compete with the usage for irrigation and rejuvenation of the Ganga. This will require consultation between the centre and the states in the federal structure of India.

The GBM basin continues to face serious water quality issues. Increasing urbanisation and associated industrialisation in the catchments of major rivers have contributed to the deteriorating situation. The three major catchments, Koshi in Nepal (a tributary of Ganga), Ganga and Padma river in Bangladesh have seen high level of degradation in water quality.[37] The role of civil society and community engagement on the Koshi-Ganga-Padma towards assessing and monitoring water quality is vital. One of the primary outcomes is to strengthen the engagement of civil society in policy-making and development initiatives to achieve the international targets on water and sanitation and improve regional co-operation between Ganga.

Reducing Mistrust

The challenges of low flow in the Ganga from the period December to March is a priority for development planning. Augmentation has benefits but will require India to cooperate with Nepal to build the proposed dams. Based on the Ganga Strategic Basin Assessment (GSBA) study, Nepal has an estimated 40,000 MW of potential hydropower in the Himalayan headwaters of the Ganga, of which only 2 per cent has been developed.[38] With a number of dams proposed, the installed capacity is likely to go up to 25,000 MW.[39] Cooperation with Nepal thus has tangible benefits; but would require managing perceptions at the diplomatic level.

In the 'neighbourhood first' approach, generosity, benevolence and kind gestures are important tools for India to build beneficial and long standing relations with its neighbouring countries. In the light of this, it is important to appreciate the Nepalese perspective and the reasons for their discomfort with water treaties signed with India. Bearing in mind the importance of having a cleaner Ganga, India should consider offering Nepal more favourable terms in order to advance future collaboration. Second, India should consider showcasing its win-win hydro-cooperation with Bhutan, despite the fact of it not having much of a reception in Nepal. Bhutan is landlocked, and blessed with abundance of water resources, much like Nepal. In 1980, Bhutan's per capita electricity consumption was 17 kWh,[40] almost the same generation capacity as Nepal. Interestingly, Bhutan's GDP per capita in 1980 was the lowest among all South Asian countries but in almost three decades thereon it had the second highest GDP per capita in South Asia – nearly 5 times that of Nepal.

It must be underlined that in terms of institutional mechanisms, India-Nepal water resource relations are built on solid structures. However, the extent of the water cooperation achieved, is below ideal levels, and is a reflection of the larger dysfunctional political relations between the two countries. The structure is as follows:

- The Joint Committee on Water Resources (JCWR); this act as an umbrella committee of various sub committees and sub groups of a certain water project. It facilitates interaction at the higher level, and is headed by Secretary, Water Resources of both countries.
- In its 3rd meeting in 2008, the JCWR recommended a three tier bilateral mechanism: Joint Ministerial level Commission on Water Resources (JMCWR) headed by the Ministers and Secretaries of Water Resources of both countries as also the existing JCWR and the Joint Standing Technical Committee (JSTC) on flood management, inundation problems, and flood forecasting.
- The structure now is: JMCWR > JCWR > JSTC.
- Another major decision of the JCWR was to set up the Joint Committee on the Koshi and Gandak Project (JCKGP), and specifically constitute a Joint Committee on Inundation and Flood Management (JCIFM). This is specifically for the Kosi and the Gandak.[41]

To rejuvenate Nepal-India collaboration, some fresh thinking and initiatives can be determined – for example, co-opting Bhutan as a third-party observer. The Joint Ministerial Commission on Water Resource (JMCWR) at the level of the Ministers of Water Resources of Nepal and India should take the lead, and build long-lasting trust between the water bureaucrats and politicians of both countries. This will involve, among other things, the transparent sharing of information from both sides, especially with regard to benefit projection in any water development projects.

With the current BJP-led NDA government and its focus on the Ganga, there is a real opportunity for effective collaboration on water development projects. Prime Minister's Modi neighbourhood approach and his visits to the neighbouring countries including to Nepal in August 2014 can provide the much needed momentum towards regional cooperation and the resultant benefits.

Ganga in India-Bangladesh Relations

India and Bangladesh have many things in common. They also share rivers. India played a prominent role in the independence of Bangladesh in 1971. However, India-Bangladesh relations have often been bitter, and at times bogged in a number of issues including the shared rivers. The 54 common rivers that India and Bangladesh share are vital for the growth, development, and the livelihood of the people. Given the growing pressure on water and the increasing impact of climate change on the water resources, a comprehensive approach to water management is imperative. India and Bangladesh have concluded only one agreement on water sharing: the 1996 Ganga Treaty.[42] While no water treaty can claim to completely remove the concerns of a lower riparian, in many ways the Ganga Treaty is a landmark one because of the inclusion of detailed water sharing arrangements in the dry season. The agreement manages the water sharing at the Farakka Barrage; however, increasingly, this has been challenged by the low flow of water in the Ganga for utilisation by both countries in the dry season.

Rivers have been central to India and Bangladesh relations. The Indo-Bangladesh Treaty of Friendship, Co-operation, and Peace signed in 1972 recognised each other's rights in the utilisation of the Ganga river, and decided to establish the Joint Rivers Commission (JRC).[43] It was a far-sighted institutional mechanism created to maintain communication and facilitate efforts to maximise the benefits from shared rivers. However, the Commission excluded the issue of sharing Ganga waters.[44] However, bilateral relations soured when the Farakka Barrage Project was commissioned by India in 1975 to divert the Ganga waters into the Bhagirathi-Hooghly river system for the preservation and maintenance of the Calcutta port.[45] For Bangladesh, the Farakka Barrage remains a sore subject – it is seen as an insensitive and deliberate action that ignores downstream concerns.

The Farakka Barrage demonstrates the difference in view between an upper riparian and a lower riparian country. By 1974, India and Bangladesh realised that during the lean season, the Ganga flow was not sufficient to maintain the Calcutta port and provide water for use in Bangladesh. Thus, while augmentation was imperative, both countries differed on the solutions. For Bangladesh, storage facilities in the upper Ganga areas in India and Nepal were viable solutions. India, on the other hand, considering its interest and concerns, recommended building canals to divert surplus water to the Ganga from the Brahmaputra – another shared river between the two countries. India stuck to the merits of the JRC and bilateral approach to water issues, and thus opposed the idea of involving Nepal. Bangladesh, as a water dependent country, feared disastrous consequences of India's structural interventions including river-linking plans.

In 1976, Bangladesh brought the case of India's unilateral withdrawal of water at the Farakka Barrage to the Islamic Foreign Ministers Conference, the Colombo Summit of the Non-Aligned Movement, and the United Nations General Assembly.[46] Following recommendation by the UN Special Political Committee, India and Bangladesh reached a five-year Agreement in 1977 in which Bangladesh was allocated 60 per cent of water volume at the Farakka Barrage, and a guarantee clause of 80 per cent of water share in case of low flows.[47] After the expiration of 1977 agreement, two MOUs in 1982 and 1985 were signed. The MoUs did not have any guarantee clause; instead it posited an equal share when the flow falls below 75 per cent of the standard flow. From 1988 to 1996, there was no formal settlement on the flow share on the Ganga waters, and during this period, Bangladesh felt aggrieved over what it considered India's unilateral withdrawal of Ganga waters.

Till the Ganga Treaty was signed in 1996, Bangladesh put water issues on top of its list for discussion with India, and domestically water became a political, emotive, and divisive issue. The Farakka Barrage became the symbol of destruction in the eyes of Bangladesh. Eventually, on 12 December 1996, the Ganga Treaty was signed in New Delhi.[48] The Treaty recognised Bangladesh's rights as a lower riparian nation, and specified the amount of water released by India at the Farakka barrage in the dry season. More specifically, the sharing of water is on 10-day periods and based on 50:50 basis if the total flow reaches 70,000 cusecs or below. The amount will be slightly varied if the flow surpasses 70,000 cusecs and 75,000 cusecs.[49]

Water watchers in South Asia often express the Ganga Treaty as a 'primitive' water sharing agreement, lacking the comprehensiveness of benefit sharing. Others view the Treaty as being limited, lacking in vision, and inherently grievance-ridden. The power-asymmetry argument of the Ganga Treaty ascribes India as maintaining its hegemonistic supremacy and *status quoist* position.[50] These observations are very pertinent, and prove the point that water sharing by nature is contestable and politically challenging. Despite the historical difficulties of water sharing and the emotions involved, the 1996 Treaty continues to function – just like the Indus Waters Treaty. Importantly, it espouses the principal of international water law – for example, the principles of 'equity', 'fairness', and 'no harm rule to either side'. Furthermore, the Treaty has given a template as well as confidence to cooperate and frame sharing arrangements on other rivers shared by India and Bangladesh.

A number of contentious issues remain. First, according to the Treaty, the amount of water allocated to India and Bangladesh at the Farakka Barrage is based on data between 1949 and 1988. As compared to the current flow, this data is a mismatch. Not surprisingly, Bangladesh alleges that India over-withdraws in the upstream tributaries, thus causing significant reduction in water flow at the Farakka. Second, the long-standing problem of augmenting the Ganga flow during the dry season will be crucial in the future for the use of both countries, and both will have to agree how best to enhance augmentation. Third, the Ganga Treaty has been the only arrangement that covers one of the 54 shared rivers between the two states. Hence, a comprehensive approach on the river and water management between India and Bangladesh is needed.

Settling water issues is paramount to a robust bilateral relation between India and Bangladesh. Misperception and misinterpretation on water is a source of anti-India sentiments in Bangladesh which extremist organisations tend to use to their advantage. India is blamed for all water woes in Bangladesh, whether it is flood or drought. The ruling government in Bangladesh is always under pressure by the opposition to effectively deal with India on water issues. India is frequently caught in the confrontation between the Bangladesh Nationalist Party (BNP) and the AL over water issues. The BNP has condemned the AL regime for signing the 1996 Treaty that allocated an unequal share of water to Bangladesh. For India, this is certainly unsettling as anti-India feelings can adversely affect bilateral ties.

Cooperation on Ganga

The 2015 resolution of land and maritime boundary disputes potentially creates space for convergences in sectors of water and energy cooperation. The traction will help frame a holistic approach to water issues including protection of the ecosystem. Both India and Bangladesh suffer water shortages during the dry season lasting from December to May. Further, population growth and increasing evidence of climate change are impacting fresh water supplies. Also, the 30-years contract of the 1996 Ganga Treaty will expire in 2026. As the upper riparian, India has responsibilities towards its lower riparian; but equally, it has its own interests to protect and safeguard. A low flow in the Ganga means diminished water supply for India as well. However, the difficulties that Bangladesh experiences as a lower riparianare, of course, far greater. Thus, augmentation of the Ganga flow is critical.

Creating cooperation with Nepal, and organising a multilateral common basin management would go a long way in helping mitigate the water dispute between India and Bangladesh. In December 2008 after being elected to power, Prime Minister Sheikh Hasina had said that she "wanted her rivers and waterways back". Likewise, India's Prime Minister Narendra Modi also believes, just as passionately as Hasina does, that rivers have to be rejuvenated and restored. The emphasis on the development of waterways is strong in the current regime of the two countries. Waterways development would connect upstream and downstream activities in terms of low cost transportation of humans and goods and significantly reducing the carbon footprint of the two countries.

Multilateral Mechanisms

Water cooperation and water governance in the Ganga basin will require a new multilateral common basin management framework, involving Nepal along with India and Bangladesh. As mentioned earlier in the chapter, Nepal is located upstream of the Ganga River, and occupies 13 per cent of the river basin. As an upstream country, one of the main proposals of building storage dams in Nepal is to improve water quality, and mitigate water shortage in downstream India and Bangladesh. The proposed dams will help regulate the lean season flow by releasing the accumulated water in the river during the dry season. The regulation and augmentation of the flow would allow the river to sustain its self-purification capability. More importantly for Bangladesh, decreasing pollution in the Ganga will improve the health and environmental conditions of the Bangladesh people where 30 per cent of the population relies on the river water for their domestic use. Regional experts have suggested 'market-based water transfers from Nepal' to augment the Ganga flow.[51] The primary objective of such an approach is not only to improve water flow but also to improve water supply for irrigation and domestic use. Such an arrangement, it is argued, will reduce Bangladesh's water insecurity, will allow Nepal to participate in the sub-regional wellness and India, as a transit country, can better manage the social and political dynamics.

It has to be remembered that the Ganga Treaty lays down that if the flows at Farakka reduce substantially due to upstream extraction, India is not under any obligation to protect the flow. With no minimum guaranteed flow for Bangladesh, water sharing becomes very charged. The 1996 Treaty also does not contain any mechanism to approach other riparian countries of the Ganga basin for the integrated management of the basin. A common basin management approach will require a fundamental shift from a bilateral emphasis to a regional outlook. In fact, at the 37th meeting of the JRC in 2010,[52] the Bangladesh delegation proposed that Article VIII of the Ganga Treaty, referring to the long-term augmentation of the flow of the Ganga, could be implemented by jointly building a reservoir at a suitable location in Nepal.[53]

This will be achieved through three complementary work streams or components that will deliver specific outputs as follows:

1. Increased awareness and experience of men and women living in river basin communities, local government and other actors on transboundary water resource management issues and policies.
2. Increased documentation of experiences and best practices in transboundary water management.
3. Strengthened capacity of men and women from river basin communities and civil society to participate in and influence transboundary water resource management decisions and/or processes.

Comprehensive Basin Management (CBM)

With new knowledge and wider riverine understanding, basin management has gained considerable traction. India has an advantage to prioritise it in the 'neighbourhood first' approach. Regional cooperation on water resources is recognised in the water policy of Bangladesh, with exchange and sharing of information and data, and joint assessment of the basins' potentials. Nepal too recognises the potential benefits for sharing water resources on equitable terms, and seeks to enhance regional cooperation in the sharing and exchange of data, as well as improving disaster forecasting and warning systems. The water policy of Bhutan expresses similar pledges with regards to regional cooperation. Regional diplomacy should emphasise sharing benefits through an integrative approach and, possibly, involve successful institutions like the UNESCO and the UNEP in the basin management approach.

Emphasis on inland river navigation which is not only cost effective but also energy efficient is an important part of basin understanding. The flow regulation through the creation of multi-purpose reservoirs in the upper reaches of the rivers will open opportunities for inland river navigation in the downstream reaches, thus benefiting all the countries in the GBM. The Ganga is National Waterway 1, and the Inland Waterways Authority of India needs to be cooperated and strengthened in basin diplomacy. The Inland Waterways Authority of India (IWAI) can play a critical role through surveys of the river bed, determining the deepest part of the channel over which boats can safely travel, and the mapping of this information. Also, these data and maps should be easily available for transporters. Smart cities on the Ganga will require modern port facilities for commerce.

In April 2016 the Indian Parliament approved the National Waterways Bill. The Bill provides for enacting a central legislation to declare 106 additional inland waterways as the national waterways in addition to five existing national waterways. The passage of the Bill highlights the crucial importance of waterways for economic development, which for long remained a backburner. Inland water transport is also a good project for regional cooperation, both in the east and west of India. The Indus river system offers great potential for inland water transport for the northern and north-western states. Likewise, the Ganga system offers for the northern and eastern states and the Brahmaputra for the eastern and north-eastern states.[54]

The experience and best practices from other river basin organisations (like the Nile Basin Initiative (NBI) and the Mekong River Commission (MRC)) can bolster efforts to frame a CBM on the Ganga. The NBI involving ten riparian countries around the Nile basin (including Egypt, Kenya and Ethiopia) began in 1999 for the purpose of dam construction and water resource sharing. Later, through negotiations, a Cooperative Framework Agreement was reached, with the Nile Council of Ministers as the highest decision making policy body.[55] The NBI plays a central role in the sustainable social and economic development in the region. The MRC, on the other hand, is an intergovernmental river basin organisation established in 1995 to deal with Mekong River issues by four lower riparian countries – Cambodia, Laos, Thailand and Vietnam. The MRC Council, the highest decision-making body, is responsible for the sustainable management of the Mekong basin.[56]

Keeping engaged with China not only on the Brahmaputra but also on the many tributaries of the Ganga which originate from the Tibetan plateau is an important aspect of the CBM. During the visit of Chinese Premier Li Keqiang to India in May 2013, serious time and discussion was given to water issues. India's proposal of a joint mechanism for better transparency on the dams being constructed on the Yarlung was diplomatically appreciated. The Chinese followed the tested 2002 MOU format, and renewed the pact on twice-daily sharing of hydrological data of the Brahmaputra River during the monsoon.

From Source to Mouth

Reliable access to sufficient quantity and quality of water for livelihood and for eco-system services are critical drivers of development. Yet for those living along the Ganga river basin in Nepal, India and Bangladesh, this is being increasingly undermined by the complex national and regional hydro-political agendas. Rights, equity or sustainability are largely ignored. Resultantly the Ganga basin is impacted by floods, prolonged water logging, droughts, significant reduction in water availability, over-extraction, water contamination, all of which are further exacerbated by erratic climatic conditions and lack of water governance systems. Further, institutions follow a top-down approach in the decision making on transboundary water issues with almost no community engagement. It is thus crucial to make transboundary water governance grounded in local contexts and be more inclusive and sustainable. Developing civic engagement models that incorporate grass-root experiences in addressing 'water insecurity' and building local capacity to engage with state authorities are important inputs to national level decision-making.

Despite the enormous potential, cooperation between the riparian states on the Ganga basin has been limited. Part of the problem lies within broader political problems. However, improved water relations can act as a catalyst for political cooperation. These include a broader inclusion of stakeholders within national discussions of water priorities, a change in the outdated mindset over hydrologic data secrecy, improved informal collaboration across borders to build trust and develop alternative thinking and options, and more efforts to de-link water issues from broader political questions. Ganga basin cooperation between India and Nepal and Bangladesh will have widespread benefits, none more significant than the livelihood of the people. Domestic political changes have opened up possibility for new diplomatic efforts and shared benefits including those related to hydropower, irrigation, flood control and salinity management. But this would require serious political give and take.[57]

As the upper riparian state, Nepal is well-positioned to help India increase the Ganga flow from Nepalese tributaries that feed the Ganga. Attention to the marginal rivers will be important, especially as these tributaries contribute more than two-thirds of river flow to the Ganga in the eastern part of India during the dry season. However, the World Bank National Ganga River Basin Project, challenges some of the common assumptions of flow management.[58] For example, on the opportunities to use storage in upstream Nepal to mitigate flooding in downstream India and Bangladesh, the World Bank report concludes that the benefits to be are much smaller than commonly assumed. While it recommends that targeted storage in Nepal could potentially increase critical low flows, it is cautious to suggest vast numbers of multi-purpose dams to maximise water benefits.

If properly considered and evaluated, storage dams in Nepal can benefit India by increasing river flow by one-third to triple the amount, depending on the month of the dry season, while Nepal stands to benefit economically from the sale of electricity generated from hydropower. India could even buy such electricity to pump underground water to augment river flow for other parts of the Ganga to further reduce pollution in the river.[59] As Nepal slowly emerges from the domestic political chaos, India should build on the goodwill generated by Prime Minister's Modi's visit to Nepal, the first by an Indian Prime Minister in 17 years, to overcome the mistrust between the two countries. A new template for furthering water relationship between

India and Nepal needs to be developed taking into account new knowledge, different approaches and local experiences to water management.

The increasing involvement of CSO can help to increase awareness and experience of people living in river basin communities, local government and other actors on transboundary water resource management issues and policies. It can also increase documentation of experiences and best practices in transboundary water management and strengthen capacity of men and women from river basin communities and civil society to participate in and influence trans-boundary water resource management decisions and/or processes.

The Ganga basin falls in what is described as South Asia's "poverty square," with substantially more people below the dollar-per-day poverty line than in all the countries of sub-Saharan Africa combined. Cooperation to improve the productive uses of Ganga waters, reduce their destructive impacts, and maintain water quality and ecosystem services should be one of the greatest engines anywhere for poverty reduction. Stability in the neighbourhood will help India focus on its development and economic growth, which in turn will have benefits for its neighbours. Addressing water-related issues through common basin management involving Nepal and Bangladesh as basin partners will help in the overall regional development.

NOTES

1. *Managing Rivers Wisely: Ganges Case Study*, World Wildlife Fund for Nature (WWF), at http://d2ouvy59p0dg6k.cloudfront.net/downloads/mrwgangacasestudy.pdf
2. Faecal coliform bacteria are the most common form of microbiological contaminates of water. The presence of faecal coliform indicates the presence of organisms that cause illness. There are acceptable limits of presence of faecal coliform in water depending on the use. Drinking water cannot contain any faecal coliform while water for swimming or bathing may contain up to 400 faecal coliform colonies/100 ml, at http://www.clemson.edu/extension/natural_resources/water/publications/fecal_coliform.html
3. A.C. Shukla and V. Asthana, *Ganga: A Water Marvel*, New Delhi: Ashish Publishing House, 1995, p.44.
4. According to World Bank Indicators, India's urban population in 2014 was 419 million, with an annual growth rate of 2.4 per cent, which is 32.4 per cent of India's total population. Please see, http://www.tradingeconomics.com/india/urban-population-percent-of-total-wb-data.html
5. See population estimates in the World Bank Ganges Strategic Basin Assessment (2012) and ICIMOD, 2007
6. In writing this chapter, the author acknowledges the interaction and inputs from the Lee Kuan Yew School of Public Policy, Governance Study Project on Ganga Rejuvenation, 2014-2015 and the Actionaid India Report, *Blues Beyond Boundaries*, 2015.

7. *Ganga Water Quality Trend*, Central Pollution Control Board, Ministry of Environment and Forests, December 2009. Please see, http://cpcb.nic.in/upload/NewItems/NewItem_168_CPCB-Ganga_Trend%20Report-Final.pdf. Also see, D. Mukherjee, M. Chattopadhyay and B.C. Lahiri, 'Water Quality of the River Ganga (The Ganges) and some of its Physico-Chemical Properties', *The Environmentalist*, 18(3), 1993, pp. 199-210; Saba Hasan, 'Water Quality of River Ganga – Pre and Post GAP: A Review', *International Journal of Advanced Research in Science, Engineering and Technology*, 2(1), January 2015; and the much cited work by Subhajyoti Das, 'Cleaning of the Ganga', *Journal of the Geological Society of India*, 78(2), August 2011, pp.124-130.
8. *Pollution Assessment: River Ganga*, Central Pollution Control Board, Ministry of Environment and Forests, Government of India, 2013, at http://cpcb.nic.in/upload/NewItems/NewItem_203_Ganga_report.pdf
9. The Ganga Basin has a catchment area of 1.08 million sq. kms that spreads across four countries: the Tibet Autonomous Region (TAR), Nepal, India, and Bangladesh. India has the largest basin coverage, with 862,000 sq. kms (roughly a quarter of India's total geographical area). See, 'Hydrology and Water Resources Information System', National Institute of Hydrology, at http://www.nih.ernet.in/rbis/basin%20maps/ganga_about.htm
10. Sarada Barrage Project Assessment between British India and Nepal 1920, p.297. Please see, http://www.internationalrivers.org/files/attached-files/treaties_between_nepal-india.pdf
11. N.D. Gulhati, *Development of Inter-State Rivers: Law and Practices in India*, New Delhi: Allied Publishers, 1992, p.166. Also see, T. Upreti, *International Watercourses Law and its Application in South Asia*, Kathmandu: Pairavi Prakashan, 2006, p. 187.
12. Sarada Barrage Project Assessment between British India and Nepal 1920 p. 297, at http://www.internationalrivers.org/files/attached-files/treaties_between_nepal-india.pdf
13. Ibid., pp. 301-308.
14. B.G. Verghese, R.R. Iyer, P.K. Ahamad, B.B. Pradhan and S.K. Malla (eds.), *Converting Water Into Wealth: Regional Co-operation in Harnessing the Eastern Himalayan Rivers*, New Delhi: Konark Publications, 1994, pp. 31-33.
15. Salman M.A. Salman and Kishor Uprety, *Conflict and Cooperation on South Asia's International Rivers: A Legal Perspective*, The Hague: Kluwer Law International, 2002, p.187
16. South Asia Water Initiative (SAWI), *Ganges Focus Area Strategy: 2013-2017,* at https://www.southasiawaterinitiative.org/ganges
17. For further information on the Joint Rivers Commission, at http://www.jrcb.gov.bd/about_jrc.html
18. M. Humayun Kabir, 'Obstacles to India-Bangladesh Cooperation', in E. Sridharan (ed.) *International Relations Theory and South Asia: Security, Political Economy, Domestic Politics, Identities, and Images,* Vol. 1, Oxford Scholarship, 2011.
19. WECS, *Water Resources in Nepal in the Context of Climate Change 2011*, Kathmandu: Water and Energy Commission Secretariat, at http://www.wecs.gov.np/uploaded/water-recource-climate-change.pdf
20. Ibid., Executive Summary.
21. Santa Bahadur Pun, 'Conflicts over the Ganga', in the volume *Disputes over the Ganga: A Look at Potential Water-Related Conflicts in South Asia,* Kathmandu: Panos Institute South Asia, 2004. Also see, Dwarika Dhungel, *Ganga Basin: Yet a Mirage?* ORF Issue Brief,

September 2013, at https://asiafoundation.org/resources/pdfs/ORFIssuebrief61DwarikaN.Dhungelformail.pdf

22. Dwarika Dhungel, *Ganga Basin: Yet a Mirage?* The author cites Mohan Lohani, 'Harnessing Nepal's Water Resources in National Interest: A Book Review', *Journal of the Nepal Council of World Affairs*, 2013.
23. Statements and Declarations of SAARC Summits of the Heads of State or Government (1985-2010), Kathmandu: Institute for Foreign Affairs, 2010, pp.1-3, at http://ifa.org.np/wp-content/uploads/2012/07/SAARC.pdf
24. Cited in Abha Dixit, *SAARC: Towards Greater Cooperation*, at http://www.idsa-india.org/an-jul-5.html
25. Arndt Michael, *India's Foreign Policy and Multilateralism*, London: Palgrave Macmillan, 2013.
26. Dhaka Declaration 1985, at http://www.saarc-sec.org/userfiles/01-Dhaka-1stSummit1985.pdf
27. Hydroelectricity Investment and Development Company Limited, at http://www.hidcl.org.np/nepal-hydropower.php
28. Dr. Babu Ram Bhattarai's statement in 2009, at http://www.telegraphnepal.com/headline/2009-12-02/sugauli-treaty-1816-a-blow-to-nepal-nationalism:-bhattarai
29. Medha Bisht, *Revisiting the Kosi Agreement: Lessons for Indo-Nepal Water Diplomacy*, IDSA, 22 September 2008, at http://www.idsa.in/idsastrategiccomments/RevisitingtheKosiAgreement_Medha%20Bisht_220908.html
30. GMR Group Press Release, 22 September 2014, at http://www.gmrgroup.in/pressreleasedetail-22-Sep-2014.aspx
31. 'Nepal allows Coastal Projects India to invest in hydropower in Nepal', *Business Standard*, 4 July 2013, at http://www.business-standard.com/article/news-ians/nepal-allows-coastal-projects-india-to-invest-in-hydropower-project-113070400561_1.html
32. SJVN Limited Press Release, at http://sjvn.nic.in/project-details.htm?13
33. 'Nepal, India and Bangladesh to make most of Ganga water, hydropower', *The Hindu*, 15 April 2013, at http://www.thehindu.com/news/international/south-asia/nepal-india-bangladesh-to-make-most-of-ganga-water-hydropower/article4617600.ece
34. The 'pre-emptive and pro-active approaches' of the CSE is well appreciated and recognised. The Governance Study Project on Ganga Rejuvenation at a presentation on 9 July 2015 titled 'Public-Private Partnerships: What would be best for Ganga', at the Lee Kuan Yew School of Public Policy in Singapore explained the CSE approach.
35. Sunita Narain and Soupamo Banerjee, *Ganga: The River, its Pollution and What We Can do to Clean It*, Centre for Science and Environment (CSE), India, 2014, at http://www.cseindia.org/userfiles/ganga-the-river-pollution.pdf
36. National Ganga River Basin Authority, *Objectives of Ganga Action Plan Phase II*, at http://www.moef.nic.in/sites/default/files/ngrba/objective2.html
37. Water Pollution Control - A Guide to the Use of Water Quality Management Principles, United Nations Environment Programme, the Water Supply & Sanitation Collaborative Council and the World Health Organisation, 1997.
38. *Ganges Strategic Basin Assessment: 2012*, World Bank, 2012. This publication established that the aggregate active water storage along the Nepalese tributaries of River Ganga was about 100 billion cubic metre (BCM), of which about half was already built (56 BCM), and half was being proposed (44 BCM). This was about 20 per cent of the 500 BCM average annual flow of Ganga. See, http://ecnphlgnajanjnkcmbpancdjoidceilk/https://

www.southasiawaterinitiative.org/sites/sawi/files/Ganges%20Strategic%20Basin%20 Assessment_A%20Discussion%20of%20Regional%20Opportunities%20and%20Risks.pdf

39. Ibid., p.59
40. Olli Varis, Cecilia Tortajada and Asit Biswas, *Management of Transboundary Rivers and Lakes*, Berlin: Springer-Verlag, 2008, p.147
41. All these mechanisms deal with projects like the 5600 MW Pancheshwar Multipurpose Project on the river Mahakali or the Sapt Kosi High Dam and the Sun Kosi Storage (SKSK) projects. The JCWR had its 5th meeting in November 2009 and its 6th meeting in November 2011. The 1st meeting of the JMCWR was held in February 2012. The 1st meeting of the JSTC was held in December 2008. The 2nd Meeting was held in March 2010. Generally, all these working groups and committees meet every two years.
42. The Treaty between Bangladesh and India on Sharing of the Ganga/Ganges Waters at Farakka, 1996. See the full text of the Treaty, at ecnphlgnajanjnkcmbpancdjoidceilk/http://www.jrcb.gov.bd/attachment/Gganges_Water_Sharing_treaty,1996.pdf
43. See the full text of the 1972 Treaty, at http://www.commonlii.org/in/other/treaties/INTSer/1972/6.html. Also see, Salman M.A. Salman and Kishor Uprety, n. 13. *Conflict and Cooperation on South Asia's International Rivers: A Legal Perspective*, p.128
44. Tauhidul Anwar Khan, 'Management and Sharing of the Ganges', *Natural Resources Journal*, 36, 1996, p. 462.
45. See, Farakka Barrage Project, at http://fbp.gov.in/
46. Mohammad Abul Kauser and Mohammad Abus Samad, 'Political History of Farakka Barrage and its Effect on Environment in Bangladesh', *Bandung: Journal of the Global South*, 2 (16), 2016. Open Access, http://bandungjournal.springeropen.com/articles/10.1186/s40728-015-0027-5
47. See the text of the 1977 Agreement, at http://ecnphlgnajanjnkcmbpancdjoidceilk/http://gis.nacse.org/tfdd/tfdddocs/394ENG.pdf
48. The Treaty between Bangladesh and India on Sharing of the Ganga/Ganges Waters at Farakka, 1996.
49. See Annexe 1, of the Treaty on Sharing of the Ganga/Ganges Waters at Farakka, at http://webcache.googleusercontent.com/search?q=cache:q7ERFO9rf-YJ:www.jrcb.gov.bd/attachment/Gganges_Water_Sharing_treaty,1996.pdf+&cd=2&hl=en&ct=clnk&gl=in
50. M.Q. Mirza, 'The Ganges Water-sharing Treaty: Risk Analysis of the Negotiated Discharge', *International Journal of Water*, 2(1), 2002, pp. 57-74. See also, R. Singh, *Transboundary Water Politics and Conflicts in South Asia: Towards Water for Peace*, Centre of Democracy and Social Action, 2008; Mohammad Khalequzzaman and Zahidul Islam, 'Success and failure of the Ganges Water-Sharing Treaty', WRE Forum, 2 November 2012, at http://wreforum.org/khaleq/blog/5689; and G. Price et al., *Attitudes of Water in South Asia*, Chatham House Report, June 2014, at http://ecnphlgnajanjnkcmbpancdjoidceilk/https://www.chathamhouse.org/sites/files/chathamhouse/field/field_document/20140627WaterSouthAsia.pdf
51. Anik Bhaduri and Edward B. Barbier, International water transfer and sharing: the case of the Ganga River, *Environment and Development Economics*, 13(1), 2008, pp. 29-51. Also see, M. Kamruzzaman, S. Beecham and G.M. Zuppi, A model for water sharing in the Ganges River Basin, *Water and Environment Journal*, 26, 2012, p. 315.
52. See, MEA, Government of India website, at http://mea.gov.in/outoging-visit-detail.htm?1122/IndiaBangladesh+37th+Joint+Rivers+Commission

53. 'FS to discuss water sharing in Delhi', *The Daily Observer*, 25 January 2016, at http://www.observerbd.com/2016/01/25/132918.php
54. Uttam Kumar Sinha, "National Waterways Bill: A Flow of Progress", *The Nationalist*, April 2016, at http://ecnphlgnajanjnkcmbpancdjoidceilk/http://spmrf.org/wordpress/wp-content/uploads/2016/04/The_Nationalist_April2016.pdf
55. See, Nile Basin Initiative website, at http://www.nilebasin.org/
56. See, Mekong River Commission website, at http://www.mrcmekong.org/
57. In my conversation with Dipak Gyawali in Colombo on 06, November 2015. Nepal's former Water Resource Minister told me that "Let us not overplay this too simplistically. There are very contradictory constraints even within a country to operate these. For instance, to maximise flood control benefits, a reservoir has to be kept as empty as possible and to maximise electricity and irrigation benefits as full as possible. How to optimise these two contradictory aims requires serious political give and take."
58. Since 2011, the World Bank has been providing financial and technical assistance to the Government of India through the National Ganga River Basin Project, at http://www.worldbank.org/en/news/feature/2015/03/23/india-the-national-ganga-river-basin-project. The role of the World Bank and the UNDP in the development of common basin management has been well documented. See, Zahir Uddin Ahmad, "Water Development Potential Within a Basin-wide Approach: GBM Issues" in Asit Biswas, Olcay Unver and Cecilia Tortajada (eds.), *Water as a Focus for Regional Development,* New Delhi: Oxford University Press, 2004, p.112. Both the World Bank and the UNDP have regularly expressed willingness to negotiate with the parties in the region for water cooperation. For the Ganga pollution problem, the World Bank in its study of 2012 recommended the construction of 23 dams in Nepal, and underlined the importance of enhanced regional cooperation in water, weather, and climatic information. The UNDP also emphasises and calls for regional cooperation in effective water governance and poverty reduction. See, *Ganges: Strategic Basin Assessment*, World Bank South Asia Regional Report, 2013, at www://ecnphlgnajanjnkcmbpancdjoidceilk/https://www.southasiawater initiative.org/sites/sawi/files/Ganges%20Strategic%20Basin%20Assessment_ A%20 Discussion%20of%20Regional%20Opportunities%20and%20Risks.pdf. Also, *Water Governance for Effective Poverty Reduction*, UNDP Report, 2004, at www://ecnphlgnajanjnkcmbpancdjoidceilk/http://www.undp.org/content/dam/aplaws/publication/en/publications/environment-energy/www-ee-library/water-governance/water-governance-for-poverty-reduction/UNDP_Water%20Governance%20for%20 Poverty%20Reduction.pdf.
59. This was suggested in the Governance Study Project on Ganga Rejuvenation at a presentation titled, 'Regional Collaboration in the Ganga Basin: Prospects for a Cleaner Ganga and Regional Peace', at the Lee Kuan Yew School of Public Policy in Singapore on 9 July 2015.

4

India-Pakistan and the Waters of the Indus

Stanley Wolpert, a noted Indologist wrote, "In 1947, India and Pakistan were born to conflict."[1] The Indian Sub-continent was partitioned in August 1947 and led to the creation of Pakistan. The partition was tragic resulting in one of the greatest forced migrations in human history. It also meant the division of the irrigation boundaries—the erstwhile highly developed and unified Indus Basin Irrigation System that irrigated some 37 million acres of land was severed.[2] Resultantly, the headwaters went to India and not unexpectedly in April 1948, a water dispute arose between India and Pakistan. Both the government thereon, desirous of attaining the satisfactory utilisation of the waters of the Indus river system, eventually concluded the Indus Waters Treaty (IWT).[3] The Treaty was signed in Karachi on 19 September 1960 by the Indian Prime Minster Jawaharlal Nehru and Pakistani President Ayub Khan.

Often the question is asked whether the IWT has served the interests of India and Pakistan. A considered, if not a prompt, response is yes. While the IWT, by no stretch of imagination, is the best treaty but in the circumstances that it was negotiated it was the most practical arrangement that the two rival countries could have signed. The IWT is not full proof and makes no claim to it. The fact that it is subjected to various stresses today is because over time growing water pressures have gripped both the countries. To address wide-ranging concerns over water between India and Pakistan, abandoning the IWT is not the solution. On the contrary, it would require careful assessment of the provisions and restrictions that form the heart of the IWT. The Treaty needs to be treated with maturity and respect and not political squabbling. Whether it requires some amendments or greater degree of transparency is a question of interpretation. But eventually, the way forward to sustaining water relations between India and Pakistan will require hydrological knowledge, technological advancement, multi-stakeholder participation and importantly apolitical response. The dilemma, however, that water is political and is largely framed within the political context persists.

The seriousness of water dispute between India and Pakistan is because of political rivalry and not because of the water issues *per se*. Political misunderstanding complicates water relations, which is well settled mutually under the IWT. The Treaty was signed to allay fears of a lower riparian and ensure Pakistan's agriculture-based economy. The IWT guarantees the water security of Pakistan, which in other words means that India as the upper riparian has to exercise responsibilities. The fact that the IWT has survived the 1965 and 1971 wars, the Kargil conflict, the worsening of relations in the wake of the Parliament attack in 2001, and the Mumbai terror strikes in 2008, and continues to function is because India has not unilaterally abandoned or disrespected it.

Despite the adversarial relations between the two countries, the IWT doggedly remains itched as a positive outcome of what has, otherwise, been a difficult bilateral relation. Some experts believe that India and Pakistan, despite their long standing hostility, are 'water-rational' states, interested in securing long-term supplies of freshwater.[4] It is highly unlikely that either of the two riparians will abrogate it, more so Pakistan, which, as a lower riparian, has virtually no choice. For India, the Treaty is a show case of its responsibilities as an upper riparian. Globally, it is difficult to find any upstream nation that assiduously abides by the principles of water sharing. It is not to make a virtue of the commitment but more to suggest the pragmatism of the hydrological reality. Better water management and greater engagement will be the new instruments of the IWT.

There is also strong opinion prevalent in India that the Treaty was generous to Pakistan.[5] Of the total water of the Indus river basin, 80 per cent flows to Pakistan via the Western Rivers, while only 20 per cent flows through the Eastern Rivers. Thus, the Treaty prescribed 80 per cent of water to Pakistan.[6] Calculations abound and many Indian experts have expressed that India should have got 42.8 per cent of the Indus basin's water.[7] Many equally feel that the IWT paid little heed to the future requirements of the population of Jammu and Kashmir. With coalition governments becoming a norm both at the Centre and the State, the views of the Kashmiri people, their aspirations, and demands remain critical to the functioning of the Treaty.

For much of IWT's history, the waters of the Indus basin system[8] flowed through the provisions defined. The odd disputes were mutually resolved by the Permanent Indus Water Commission (PIC).[9] However, since the rise of terrorism in Jammu and Kashmir in 1989, water has gained political ascendancy, often testing the durability of the IWT. The public discourse in Pakistan on the water sharing have been strident and the 'fairness' argument, after so many decades, continues. Upper riparian states respond to their hydrological position by maximising the use of water in their catchment as a sovereign right and national interest. Lower riparian fear unilateral withdrawals and seek negotiations over water sharing. While the IWT permits India to utilise 1.50 MAF of water from Jhelum for its own use and the right to construct run-of-river multi-purpose projects, the question of how much water is being diverted and in which season becomes worrisome for a down riparian. These factors leave Pakistan feeling vulnerable to upstream water use.[10]

Legacy of IWT[11]

There is a large number of work that comprehensively deal with the history of the IWT.[12] ND Gulhati's first-hand account of the negotiation process is an important source of information. As the Indian chief negotiator, Gulhati's work covers in great detail the fears and apprehensions of Pakistan, the water needs of India, and the role of the World Bank.[13] Below is a summary of the important phases of water relations between India and Pakistan:[14]

Negotiations 1947-51

Undivided India had an extensive network of canals, which was supplied by the waters from the Sutlej, the Ravi, and the Beas rivers. The partition and the division of Punjab (East and West) by the Radcliffe Award, the sharing of the waters of the Sutlej, the Ravi, and the Beas became an issue that urgently needed to be resolved. An initial 'Standstill Agreement' allowed for *status quo*, until a final settlement was signed on 20 December 1947 by the East and West Punjab governments to continue with the flow of the waters. The agreement expired on 31 March 1948.

On 1 April 1948, after having notified the West Punjab of its action, East Punjab stopped the flow of water to the two canals – the Dipalpur Canal and Upper Bari Doab Canal (UBDC) in West Punjab – thereby affecting water supply in several places, including the city of Lahore. The dispute, referred to as the 'canal water dispute', was in a larger sense about the rights of the lower and upper riparian, and about equitable distribution. Gulhati describes in his book that in a telegram dated 15 April 1948, the Prime Minister of Pakistan, Liaquat Ali Khan, requested the Prime Minister of India to 'take immediate action for restoring water supply', adding 'I regret that before we have had time enough to settle our existing problems, the Government of East Punjab has thought it fit to create new ones...'[15] Gulhati further mentions that East Punjab was accused by Pakistan of 'Machiavellian duplicity.' Several motives can be attributed to India's action. One interpretation suggests that the singular motive of stopping the supply of water to the two canals for a few days, was to establish India's sovereign rights on the UBDC.

Soon after, an interim agreement, known as the Inter-Dominion or Delhi Agreement, was signed on 4 May 1948. Accordingly, waters would continue to flow from India in return for payment by Pakistan.[16] The agreement also established India's rights on the rivers in East Punjab.[17] Not happy with the outcome, Pakistan, on 16 June 1949, offered to take the dispute to the ICJ, which India refused. This was the first such move to seek arbitration, and indicated strongly that Pakistan would, in due course, internationalise the issue and seek third party mediation. As a counter response, on 10 May 1950, India registered the Delhi Agreement with the UN, while Pakistan disclaimed its validity.[18]

Between May 1950 and May 1952, Pakistan's accusation and India's counter claims stalled any progress towards a final settlement of water sharing. Not surprisingly, Pakistan even considered taking the issue to the UN Security Council.[19] During the time, David Lilienthal (former Chairman of Tennessee Valley Authority and also of the US Atomic Energy Commission), on a visit to India and Pakistan wrote publicly that the animosity between the two countries could be reduced if a joint programme of developing and operating the Indus basin river system was worked out. He suggested to the International Bank for Reconstruction and

Development, popularly called the World Bank, to extend financial help.[20] Lilienthal's proposal received support from the US State Department as well as the World Bank and, in September 1951, Eugene Black, the President of the World Bank, wrote to the Prime Ministers of India and Pakistan offering the Bank's good offices to facilitate a proposal on the lines suggested by Lilienthal. The three principles that underlined the proposal were: the Indus basin water resources were sufficient to meet all the existing as well as the further uses of both countries; the water resources of the Indus basin should be cooperatively developed by treating the Indus basin as a unit; and that the problem of the Indus basin water resources should be resolved on a functional basis and not on a political plane.[21] The Bank's proposal was to be taken as a basis, not of settlement, but for negotiation.

Negotiations: 1952-56

On 7 May 1952, a Joint Working Party of Engineers held its first meeting at the World Bank in Washington followed by another meeting in November 1952 to determine the allocation of the rivers. India proposed allocation of all the eastern rivers (the Ravi, the Beas, and the Sutlej) and 7 per cent of the western rivers (the Indus, the Chenab and the Jhelum). The earlier Pakistani plan had proposed 30 per cent of the waters of the eastern rivers to India, and none from the western rivers. Such distributional differences led to an expected impasse.[22] Some of the crucial differences are chronicled below.[23]

To bridge the 'distributional differences', the World Bank on 5 February 1954 proposed the allocation of the entire western rivers to Pakistan, and the entire flow of the eastern rivers to India. It also provided for a transition period to complete the construction of the link canals needed in Pakistan to make use of the waters of the western rivers. The proposal recommended that the cost of such works would be borne by the benefiting countries.[24] On 22 March 1954, Nehru accepted the World Bank's proposal. India's decision was influenced by the fact that it intended to open the Bhakra canal in June 1954. Pakistan, however, rejected the proposal. In his reply to the World Bank on 14 May 1954, Pakistan's Prime Minister Mohammad Ali said that its proposal 'does not in fact meet the test of fairness'. He said that it was 'neither practicable nor equitable to cut off Pakistan's historic supplies from the eastern rivers'.[25] Sensing the resistance and thus lack of progress, the World Bank on 21 May 1954, submitted a memorandum to

Pakistan suggesting some favourable adjustments. However, Pakistan continued to take an uncompromising position.

Losing patience and possibly annoyed with Pakistan's stubbornness, India, on 21 June 1954, sent a letter to the World Bank President expressing its inability to continue with the talks. This was done in view of the impending opening of the Bhakra Nangal canal. On 8 July 1954, the Bhakra canals opened and not surprisingly Pakistan immediately condemned it.[26] There was massive anti-India reaction in Pakistan and the Prime Minister of Pakistan wrote to Nehru protesting against the diversion of the Sutlej waters to Bhakra Nangal canal as a 'serious detriment of supplies to Pakistan'.[27]

Pakistan's position changed suddenly and, on 28 July 1954, the Foreign Minister of Pakistan wrote to the World Bank accepting the proposal. It remains largely unexplained in Gulhati's book as to what prompted or forced Pakistan to quickly change its stance. Possibly its earlier position was one of posturing or probably it was simply the growing reality of Pakistan's down-riparian position. Consequently, on 6 December 1954, a new round of discussions began in Washington. While a comprehensive settlement of the sharing of the Indus river basin was being worked out, Pakistan required the waters for irrigation purposes in the transition period. In the absence of a negotiated agreement, a series of transitional agreements were arrived at with a view to assuring water supplies for the seasonal crops of *rabi* and *kharif*.[28]

Final negotiations: 1956-1960

An interesting point that is revealed by Gulhati is the World Bank's sensitivity towards Pakistan as a downstream country. In order to seek Pakistan's cooperation, the World Bank came out with an *aide memoir* on 21 May 1956 suggesting an adjustment to its earlier proposal. The essential point of the Bank's memorandum was that Pakistan would be able to construct some storage on the western rivers to deal with the problem of water shortage, and that India's financial liability would be to the extent of keeping such storage to the minimum. In other words, India would have to contribute not just towards making the replacement canals but also to making storage on the western rivers which Pakistan wanted.[29]

The negotiations yet again went into a limbo and after some uncertainty, on 22 December 1958, Pakistan unconditionally accepted the Bank's proposal. This happened soon after the military coup in Pakistan. In his biography, *Friends not Masters*, Ayub Khannoted that the Pakistani approach was tantamount to 'asking for the moon', suggesting that the leadership should wake up to the riparian reality. The other contributing factor could have been India's proposed diversion of the Chenab waters at Marhu. By then the World Bank had received assurances of substantial monetary support for financing the replacement works in Pakistan[30] and this clearly would have influenced Pakistani leaders. There was much to gain for Pakistan by accepting the proposal.

The final negotiations towards the Treaty began in May 1959, and took 15 months and on 19 September 1960, the IWT was finally signed.

Role of the World Bank

It is said that an international treaty that gives one party all that it wants cannot be a good treaty. Like all treaties, the IWT was a compromise. As an upper riparian, India could have chosen not to sign the Treaty and use the waters unilaterally. In fact, in 1947, Nehru had said that India would use its water as it pleased. There were, however, compulsions that India could not ignore. Had the Treaty not been signed, much of the socio-economic development plans in India and Pakistan would have been stymied, and tensions over water sharing would have continued to spiral the relationship into uncertainty.

It cannot be denied that the signing of the Treaty became possible because of the timely intervention of the World Bank, and subtle pressure from the USA. Concerned over the possibility of escalation, the serious repercussions on the economic well-being of the two countries, and keeping in mind its own institutional interests, the World Bank played a stellar role. Both India and Pakistan desperately required financial assistance for developmental projects, and the World Bank found itself in an influential position. Even before the negotiations began, the World Bank had told India to resolve its canal water dispute with Pakistan before it financed the Bhakra Nangal project.[31] The World Bank Chief Negotiator Sir William Illif remained in constant touch with the US State Department.[32] While the IWT partitioned the Indus water system, Pakistan required roughly a billion dollars to finance the construction of its link canals and storages. India was made to pay about 62 million pounds for new construction works in Pakistan, while the World Bank and the US (along with other aid-giving countries) mobilised the rest. The proposed system of canals in Pakistan was of such magnitude that, without adequate finances, the final settlement would not have been possible. Some observers even feel that the USA used its financial muscle for making the deal possible.[33]

Reactions towards the IWT

In the Lok Sabha debate on the IWT, Prime Minister of India Jawaharlal Nehru, on 30 November, said, "We purchased a settlement, if you like; we purchased peace to that extent and it is good for both countries".[34] Other Members of Parliament including those belonging to the Congress, Praja Socialist Party (PSP) and Jana Sangh raised concerns over the loopholes of the Treaty. Iqbal Singh and H.C. Mathur, Congress MPs from Punjab and Rajasthan, called the treaty disadvantageous to India stating that both their home states "had been badly let down".[35] Ashok Guha, another Congress MP lamented that "interests of India had been sacrificed to placate Pakistan". Ashok Mehta, leader of the PSP in the Lok Sabha described it as a "peculiar treaty under which Pakistan, already a surplus area, would be unable to make full use of her share of the Indus Water and would have to allow it to flow into the sea."[36]

While India has learnt to live with the Treaty and abide by its restrictions and provisions, in Pakistan the reactions become successively sharp. Various actors in Pakistan have pushed water onto the political arena so as to 'have [it] accepted by a sufficient audience to sanction extraordinary defensive moves'.[37] Syed Salahuddin, Chairman of the United Jihad Council[38] was quoted (*Ausaf* 18 June 2002) as saying, 'Kashmir is the source from where all of Pakistan's water resources originate. If Pakistan loses its battle against India, it will become a desert.'[39] A few months later, Sardar Mohamad Anwar Khan, President of Kashmir under Pakistani Control, said, 'Pakistanis who believe that they can survive without Kashmir are wrong. The Pakistani economy is dependent on agriculture and hence on water, and therefore on Kashmir.'[40] Prime Minister, Sardar Sikandar Hayat reiterated the point: 'The freedom fighters of Kashmir are, in reality, fighting for Pakistan's water security and have prevented India from constructing a dam on the Wullar barrage.'[41] Correspondingly, a senior officer of the army, Lieutenant General Zarar Azim, the Corps commander of Lahore in 2003, said, 'Kashmir is our lifeline, and its importance increases in view of our water quality.'[42] Even Musharraf strongly believed that the issue of Kashmir and the distribution of the Indus waters were interconnected. Any lasting peace, to Musharraf, would have to be based on the fair distribution of the river waters from the Pakistani perspective.[43]

Did India compromise?

This is a question that crops up even today especially in times of heightened tensions. Water was critical for India's development plans, irrigation facilities, and power generation. Thus, it was crucial to get the waters of the eastern rivers for the proposed Rajasthan canal and the Bhakra dam. Without these waters, both Punjab and Rajasthan would be left dry, and would have severely hampered India's food production.

Clearly, due to its strategic location and importance, the Indus basin attracted a great deal of seriousness and western attention. The Indus tributaries passed through Jammu and Kashmir which had received, by then, considerable international attention. In fact, after visiting India and Pakistan, David Lilienthal wrote (in the US magazine *Collier's*, August 1951) that the two countries were on the edge of a war over Kashmir, and that the US might be drawn into it. Lilienthal had feared 'another Korea ... in the making'.[44]

In India, as stated, the IWT has been perceived as being highly generous towards Pakistan for having given away 80 percent of Indus river waters to Pakistan; in Pakistan, the view has been radically different. The main impression in Pakistan has been that the loss of the eastern rivers was irreparable. Commentators like Bashir Malik have challenged the treaty provisions, saying that it was Nehru who manipulated the Radcliffe Award to ensure that the headworks of Ferozepur remained in India.[45] Malik grieves that the signing of the Standstill Agreement and the Delhi Agreement was a colossal error which, in the end, cost Pakistan its rights over the eastern rivers. He goes on to say that India's negotiation tactics were superior to those of Pakistan.[46] He also questions the World Bank's motives behind the 1954 plan, believing that it was well aware that the loss of eastern rivers would be 'a rude shock to bear with (for) Pakistan'. Malik writes: 'It would seem as a tactical strategy to assure her, though falsely, of availability of enough flow of waters of Western Rivers'.[47] He adds that the Bank's proposal 'incorporated the core elements of the Indian plan. In fact, she gained much more than she could ever imagine... She got away with the total flow of 33 maf "virtually for a song"'.[48]

The Treaty

The IWT is a technical treaty which partitions the rivers of the Indus basin. The treaty's preamble, 12 articles and 8 technical annexures (A to H), lay down in great detail the responsibilities and obligations for both the parties. The preamble states that the treaty seeks to fix and delimit the 'rights and obligations of each in relation to the other concerning the use of these waters'.[49] The treaty divided the rivers without taking the volume of water into account, and made no provisions for joint management. The treaty has no exit clause; however, it can be modified through a mutual agreement. Though an upper riparian, India has certain responsibilities with regard to the use of the waters as they are explained in articles II, III and IV.[50]

Notwithstanding the agreed water sharing formula, the IWT has seen many ups and downs in the last 50 years. The Permanent Indus Commission has held 107 meetings and undertaken 114 tours by March 2012 to resolve many outstanding issues.[51] Although Pakistan has used the threat of invoking the provisions relating to the settlement of differences and disputes on several projects, it has so far referred only one issue (on Baglihar) to a neutral expert.[52] It has served notice of intention to refer issues to a neutral expert in the Kishenganga and Nimoo Bazgo projects, but has so far not acted in this regard. Pakistan has referred those aspects of Kishenganga that it feels require interpretation of the Treaty to a Court of Arbitration, since set up under the provisions of the Treaty.

Apart from the data supplied by India on various projects in accordance with the Treaty, Pakistan has also used the provisions of the Treaty to seek data on various other projects, many of which are yet to even come up and be approved.[53] There is an impression in India that Pakistan takes recourse to the provisions relating to its right to raise objections, and seeks additional data and information for their resolution, in order to delay India's projects. There have been specific instances where the Treaty came under strain:

- In 1966-67, Pakistan complained that India was interfering with the flow of waters to Pakistan in contravention to the provisions relating to the transition period. It appears to have died a natural death.[54]
- In 1974, Pakistan objected to the Indian proposal (submitted in 1968) to build the run-of-the-river Salal Dam Project on the river Chenab. After protracted negotiations, an agreement was signed in 1978. India made changes in the design of the dam by lowering its

height to satisfy Pakistan. The dam faced severe siltation problems later. The power generation capacity of the dam reduced significantly due to the changes in design made by India.[55]

- In 1986, Pakistan objected to India's move to build a small storage-cum-navigational facility on the Wullar Lake in Jammu and Kashmir to improve navigation in the Jhelum River. This would have also assured a regular supply of water to Pakistan's Mangla dam downstream. The matter was referred to the two governments in 1986. India stopped working on the project from 1987. The project has still not been completed although it would benefit both India and Pakistan. The Tulbul Navigation Project is now a matter of discussion at the government level, and is outside the scope of the Permanent Indus Commission.
- Pakistan was informed about the construction of a dam at Baglihar on the Chenab in 1992. It objected to the design of the dam, and the discussions between the two Indus Commissioners could not resolve the differences. Pakistan invoked the IWT provision of referring the matter to a Neutral Expert in 2005. The World Bank appointed Raymond Laffitte, a Swiss civil engineer and a neutral expert, in May 2005. Laffitte gave his findings in February 2007 after visiting the dam site, talks with both parties, and analysing about 13,000 dams across the world. He suggested some minor modifications in the design which India readily accepted. But Pakistan was dissatisfied. The findings of this Neutral Expert, in India's view, served as a guideline for the design of future dams. There has been a lot of criticism in Pakistan of the government's handling of the Baglihar dispute. Baglihar was a bad experience for Pakistan; but it keeps the water issue alive in the public domain by accusing India of 'stealing' its water by reducing the flow of the Chenab.[56]

Water a Political Issue

The submission of a 'non paper'[57] on water-related issues by Pakistan during the Foreign-Secretary level talks in February 2010 added a new set of dynamics to its overall political relationship with India. At one level, the non-paper gives the impression that Pakistan is taking positive steps to energise the IWT. However, in actuality, it is pointing an accusing finger at

India. The concerns raised, primarily relating to delays in information on projects from the Indian side is simply to delay the hydel-projects. Water issues are being politically constructed in Pakistan, and its water scarcity is increasingly couched in the language of security vis-à-vis India, the upper riparian state.[58]

A new phase in Pakistan's water hostilities with India seems to have emerged, particularly after President Asif Ali Zardari's article in the *Washington Post* (28 January 2009) in which he warned, 'The water crisis in Pakistan is directly linked to relations with India'.[59] Pakistan wants to reframe a new set of lower-upper riparian dynamics by articulating its 'water rights' under the provisions of the Treaty by raising concerns (as the non-paper does), and then asking India for explanations.

The trajectory of the public discourse on water in Pakistan is such that water is increasingly being projected as a flashpoint. Given Pakistan's asymmetry in terms of hydrology, as well as economic and military resources, it will strive for a more equitable distribution of waters with India. Likewise, it will continue its proxy war, hoping to force India to negotiate to its advantage on the water issue. On its part, India is primarily concerned with state-sponsored terrorism, and will only show willingness to 'talk' about 'water needs' with Pakistan and not 'negotiate' on 'water rights'.

Jammu and Kashmir Factor

The discourse on the IWT often misses the Jammu & Kashmir factor. The three western rivers – Indus, Jhelum and Chenab – flow through J&K before entering Pakistan Occupied Kashmir (PoK). The people of J&K for long have perceived the IWT as unfair. Nehru was mindful of the needs and requirements of the J&K people.[60] During the negotiations, India had also expressed its concerns over the construction of the Mangla Dam by Pakistan in PoK, and stated that the execution of the Mangla Dam was an effort to exploit 'the territory to the disadvantage of the people of the state, and for the benefit of the people of Pakistan.'[61] Many decades later in interview in 2006, the Indian Minister of Water Resources, Saifuddin Soz, stated that the Treaty had taken care to safeguard India's interest, particularly in J&K.[62]

However, perception exists that India's generosity or rather Nehru's desire to 'purchase peace' cost the Kashmiris dearly.[63] Countering Nehru's approach, Riyaz Punjabi writes that the treaty could not buy peace as the 1965 war demonstrated, but in the bargain, 'genuine economic interests of J&K state' were surrendered.[64] There is also an argument that the Indian projection of J&K future irrigation and hydel requirement was not sufficiently determined.[65] Gulhati notes,

> None of us had, at that time, any real idea of the quantum of future developments in the upper reaches of the western rivers. Nor did we have any idea of the irrigation from the Indus in Ladakh. As regards hydro-electric development we felt that, being a non-consumptive use, it was not covered by the Bank proposal which dealt only with irrigation uses.[66]

Bashir Malik's account of the negotiation process reveals that Pakistan pitched its demand high after February 1957, when Shurawardy became Prime Minister. It expressed its unwillingness to accommodate 'any storage of water on River Chenab', and held that all 'uses for water in the Kashmir state must be quantitatively fixed'. Pakistan always feared that India would restrict flow of water into Pakistan, forcing the people to 'starve out' and 'die of thirst'.[67]

J&K is a key factor in the water debate. The population in J&K has increased three times since the signing of the treaty. This has added enormous pressure on the agriculture sector. In spite of the vast hydel potential,[68] the state has remained industrially backward. The twin issues of water and power shortages have scared away industrialists and investors, leading to unemployment, 'which in turn provides recruits for terrorism'.[69]

On 3 March 2003, the J&K Legislative Assembly in a unanimous resolution called for a review of the treaty. Many commentators in J&K have hinted at the scrapping of the old treaty,[70] and some of them have cited instances of the renegotiation of treaties at the international level.[71] The people of PoK have also raised their objections to the IWT arguing that both India and Pakistan 'have failed to incorporate the right of the people of Kashmir in the management of water uses and water-related activities under the Indus Water Treaty.'[72]

The Future of IWT

Given the political uncertainty, water sharing between India and Pakistan always draws fear of conflict, even though the IWT continues to function. In order to build the trajectory of conflict and cooperation, this section identifies six critical drivers. These are:

- *Pakistan's attitude*: Pakistan continues to be suspicious of India in so far as the implementation of the IWT is concerned, and questions the 'fairness' of the Treaty by taking recourse to the clause which provides for a neutral expert and the court of arbitration on different projects.
- *India's attitude*: India regards the treaty as 'fair' and 'generous'; but Indian public opinion has been hardening on Pakistan – particularly now with our greater sensitivity to domestic water needs. The issue of India conceding 80 per cent of the waters under the Treaty fails to reconcile with what the outcome of the 63 years of relationship has been with Pakistan. Constantly questioning India's projects on the western rivers through arbitration may prompt India to take a hard-line position on the Treaty.
- *Political situation*: The fluctuating fortunes of India-Pakistan relations can have a major impact on the functioning of the treaty. Each time there is a precipitous fall in the relationship, as has happened since the Mumbai attacks, the pressure on the Treaty to either be 'reviewed' or even 'abrogated' will mount.
- *Kashmiri viewpoint*: The people in J&K are becoming increasingly vocal in their criticism of the Treaty. Voices have been raised in the state favouring abrogation. With a population that has grown three times since the signing of the Treaty the demands and expectations are exponentially high.
- *International opinion*: How the international community perceives India and Pakistan on the sharing of the Indus water system is also an important driver because the World Bank is involved in the Treaty's functioning, and has institutional interest in de-escalating tensions. Given the geo-strategic importance and climate change vulnerability of the region, the World Bank would like to use its good offices to re-work on devising a new formula for the quantitative settlement of the shared waters.
- *Climatic factors*: There is scientific evidence that climatic factors are impacting the flow of the rivers. The reduction in flows often leads to allegations that India is stealing Pakistani waters. Floods in the rivers also lead to allegations that India is deliberately flooding the other side. Inadequate understanding of climatic factors can lead to misunderstandings and misperceptions.

Based on the above drivers, water sharing—given its political, emotional and divisive texture—can either become intensely conflictual, or the benefits accruing from the principles of water sharing can act as a catalyst for strengthening further cooperation.

Conclusion

The IWT remarkably balanced the water rights of Pakistan with the needs of India without compromising on the historical usage. While allegations by Pakistan that India has violated the provisions of the IWT abound, much of the criticism appear to be motivated. Pakistan is unlikely to get such generous terms should there be a renegotiation of the Treaty. The competition today for the waters in the Indus basin is many times more than what it was in the 1950s and, therefore, claims to the Indus waters will only become magnified on either side. The fundamental challenge, therefore, is to keep the differences within the framework of the Treaty, and evolve a mechanism of finding solutions to the immediate and many unforeseen water-related issues.

Public opinion on both sides are critical of the IWT but officially neither Pakistan nor India have communicated any desire to modify the treaty. A section of public opinion in India, particularly in response to Pakistan-sponsored terrorism, argues for the abrogation of the treaty. The IWT cannot be abrogated unilaterally. However, India may at some stage consider taking 'counter-measures' against Pakistan for not fulfilling its obligations (of not supporting terrorism) under international law and thereby contemplate abrogating the treaty unilaterally.[73] Interestingly, Article XII of the treaty says that it 'may from time to time be modified by a duly ratified treaty concluded for that purpose between the two governments'.

While the IWT offers a detailed format of provisions and restrictions on the Indus river system, there is also an urgent need to respond to the future water challenges that does not fall within the ambit of the treaty. A space for water cooperation, beyond the IWT, has to be created based on sharing new hydrological knowledge, experiences and best practices on transboundary water issues. There is a need to have updated information about the environmental flows of the Indus River System, entry of effluents and seepage losses in lakes and reservoirs. Any revision to the Treaty would need to evolve a joint mechanism that is well-supported by high quality data and analysis on water quantity, quality and identified risks and opportunities in consensus between India and Pakistan.[74]

For Pakistan it is important to focus on its domestic water management policies as well as the inter-provincial water dispute between Punjab and Sindh rather than aggressively accusing India of 'stealing' waters, which benefits the political-military class by drawing international attention. International water experts like John Briscoe, Gordon McKay Professor of Environmental Engineering, Harvard University, tend to take a sympathetic view of Pakistan ignoring the accommodation of India as an upper riparian. Briscoe's article 'War and Peace on the Indus', published in *South Asian Idea*[75] puts the onus on India as the regional hegemon, to show restraint on the Indus basin.

The Possible Way Ahead

Notwithstanding the projection of water as flashpoint, there is far greater value in sharing the benefits of water. This needs to be structured in the bilateral relations. In addition to surface water issues, ground water abstraction is a matter of serious concern. Pakistan and India share a continuous water aquifer which is not clearly demarcated between the two countries. Over-abstraction of groundwater in Punjab (Pakistan and India) is affecting the aquifer's water quality and quantity. At one level, there is hardly any bilateral engagement on the aquifer and the knowledge base in terms of data and studies are insufficient to know the current rate of abstraction.

Remarkably, the means to overcome some of the predicted water woes and energy crisis between India and Pakistan are, far-sightedly enough, laid out in the IWT itself. Article VII is about 'Future Cooperation'; it opens up a range of possibilities through the 'optimum development of the rivers' by 'mutual agreement to the fullest possible extent'. It relates to 'installing hydrological observation stations', and 'carrying out such new drainage works as may be required…' It also states that '…the two parties may, by mutual agreement, co-operate in undertaking engineering works on the rivers'. However, if such engineering work 'affects the other party materially, it shall notify the other party of its plans and shall supply such data relating to the work as may be available…'[76] Keeping in mind the 'optimum development', new dams should be selected in ways that take into account the 'health of the rivers', that includes ecological considerations, sediment loads, and flow regimes. Often, the social and ecological costs are not fully considered. A completely new orientation to dams needs to be developed, involving greater public participation.

Dialogue remains essential for water development. The public discourse on water issues between India and Pakistan is far too narrow, and largely based on misperception. Transparency will help in clearing the air, allowing for shared benefits on the waters, and building ideas of 'water peace' rather than 'water wars'. The negotiators of the Treaty were visionary, and had initially approached the dispute through ideas of joint development. The sharing of data and joint research on climate change mitigation, along with the joint development of the vast hydroelectric and irrigation potential of the western rivers for mutual benefit should be the thrust towards the new era in India and Pakistan water relations.

NOTES

1. Stanley Wolpert, *India and Pakistan: Continued Conflict or Cooperation?*, University of California Press, Berkeley, 2011, p.126.
2. M. Yunus Khan, 'Boundary Water Conflict between India and Pakistan', *Water International,* 15 (4), 1990, pp. 195–199.
3. The IWT divided the use of rivers and canals between the two countries. Pakistan obtained exclusive rights for the three western rivers, namely the Indus, Jehlum and Chenab. India retained rights to the three eastern rivers, namely the Ravi, Beas and Sutluj. The treaty also guaranteed ten years of uninterrupted water supply. During this period, Pakistan was to build huge dams, financed partly by long-term World Bank loans and partly by compensation money from India. Three multipurpose dams, the Warsak, Mangla and Tarbela, were built. A system of eight link canals was also built, and the remodelling of existing canals was carried out. Five barrages and a gated siphon were also constructed under this treaty. For readings on the negotiations that led to the IWT please see, David Lilienthal, 'Another Korea in the making', *Colliers Magazine*, 4 August 1951; A.N. Khosla, 'Development of the Indus River System: An Engineering Approach', *India Quarterly*, vol. 14(3), July-Sept 1958, pp. 234-253; Sisir Gupta, 'The Indus Water Treaty', *Foreign Affairs Report*, vol. 9 (12), 1960; N.D. Gulhati, *Indus Water Treaty: An Exercise in International Mediation*, Bombay: Allied Publishers, 1973; Asit Biswas, 'Indus Water Treaty: The Negotiating Process', *Water International*, vol. 17(4), 1992.
4. Undala Z. Alam, *Water Rationality: Mediating the Indus Water Treaty,* September 1998, at http://www.ppl.nl/bibliographies/water/files/4613.pdf. According to the water rationality thesis, countries will prefer cooperation to conflict to promote long-term security of their water supplies. p.24.
5. For a critique of the Treaty, see K. Warikoo, 'Indus Water Treaty: View from Kashmir', in *Himalayan and Central Asian Studies*, vol. 9, no.3, July-Sept. 2005, pp.14-18.
6. Many in Pakistan feel that the territories that went to India after Partition were historically 'using less than 10 per cent of the Indus waters and that the Treaty was generous to India

in giving it 20 per cent of the waters', Ramaswamy Iyer, Stimson Center Paper, Washington DC, 27 October 2008.

7. K. Warikoo, in *Himalayan and Central Asian Studies*, vol. 9, no.3, July-Sept. 2005, p.17.
8. The Indus Basin is an important geophysical part of the Indian Subcontinent. The Indus, together with the Chenab, Ravi, Sutlej, Jhelum, Beas and the extinct Sarasvati constitutes the basin. The basin has a total area of 11,65,500 kms, with annual available waters of 207 BCM. The basin countries are: Pakistan (632,954 sq km); India (374,887 km2); China (86,432 km2); and Afghanistan (76,542 sq km).
9. Article VIII of the IWT mentions the Permanent Indus Commission. Commissioners meet regularly and communicate directly and monitor the river development. See, Neda A. Zawahri, 'Designing River Commissions To Implement Treaties and Manage Water Disputes: The Story of the Joint Water Committee and Permanent Indus Commission', *Water International,* 33(4) December 2008, pp. 464–474.
10. Ramaswamy R. Iyer, 'Was the Indus Waters Treaty in Trouble?' in *Economic and Political Weekly,* June 22, 2002, pp. 2401-2402.
11. The chapter hereon gives a historical account of the partitioning of rivers and the negotiation process leading up to the signing of the Indus Waters Treaty. The historical narrative is taken from the author's previous works Uttam Kumar Sinha, Arvind Gupta and Ashok Behuria, "Will the Indus Waters Treaty Survive', *Strategic Analysis*, vol. 36, no. 5, Sept-Oct, 2012, pp. 735-42; Uttam Kumar Sinha, '50 Years of the Indus Waters Treaty: An Evaluation', *Strategic Analysis*, vol. 34, no. 5, Sept-Oct, 2010, pp. 667-670 and Uttam Kumar Sinha, 'Water a Pre-eminent Political Issue between India and Pakistan', *Strategic Analysis*, vol. 34, no.4, July-Aug, 2010, pp.482-85.
12. Aloys Arthur Michel, *The Indus Rivers: A Study of the Effects of Partition*, Yale University Press: New Haven, 1967; Scott Barrett, 'Conflict and Cooperation in Managing International Water Resources', Policy Research Working Paper 1303, The World Bank, May 1994; and Bashir A. Malik, *Indus Water Treaty in Retrospect*, Lahore: Brite Books, 2005.
13. Niranjan D. Gulhati, *The Indus Waters Treaty: An Exercise in International Mediation*, Bombay: Allied Publishers, 1973.
14. This has been culled out from the author's earlier work on the IWT. See Uttam Kumar Sinha, et al, 'Will the Indus Waters Treaty Survive?', *Strategic Analysis*, 36(5), 2012, pp.736-39.
15. Gulhati, n.13, p.63.
16. For the text of the agreement, see N.D. Gulhati, *The Indus Waters Treaty: An Exercise in International Mediation*, Bombay: Allied Publishers, 1973, p.69. The West Punjab government agreed to deposit an *ad hoc* sum in the Reserve Bank, to be specified by the Prime Minister of India, for the continued flow of waters to West Punjab through the Central Bari Doab and the Dipalpur canals in West Punjab. The contention of the East Punjab government was that the West Punjab government could not claim any share of these waters as a right under the Punjab Partition (apportionment of assets and liabilities) order 1947.
17. J.L. Nehru, *India's Foreign Policy*, Publications Division, Government of India, New Delhi, 1961, p. 477. Nehru was explaining the canal water dispute between India and Pakistan on the occasion of the opening of the Nangal Canal on July 8, 1954. Nehru complained that the agreement arrived at in 1948 was not fully complied with by Pakistan. In fact, Pakistan had denounced the agreement. He also explained how the World Bank got

involved in mediating between India and Pakistan. The 1948 agreement had bound India to only gradually increase its off-take. The agreement was meant to be of a temporary nature. But Pakistan procrastinated. Pakistan also rejected the various proposals put forward by India and the World Bank in 1954. India decided to resume its freedom of action on the use of waters of the rivers in East Punjab. Eventually, India's action spurred Pakistan to come to the negotiating table.

18. The historical details have been accounted in Uttam Kumar Sinha, et al, 'Will the Indus Waters Treaty Survive?', n.14.
19. N.D. Gulhati, *The Indus Waters Treaty: An Exercise in International Mediation*, n.16, pp.88-89.
20. Ibid., p. 94.
21. Ibid., p. 97.
22. Ibid., pp. 134-135.
23. The details have been expressed in Uttam Kumar Sinha, et al, "Will the Indus Waters Treaty Survive?", *Strategic Analysis*, 36(5), 2012, p.738-42.
24. Gulhati, n.19, p. 137.
25. Ibid., p. 151.
26. Ibid., p. 161-162.
27. Ibid., p. 163.
28. Ibid., p. 191.
29. Ibid., p. 204.
30. Ibid., p.253.
31. Undal Z. Alam, n.1, p. 231.
32. David R. Stone, 'The United States and the Negotiations of the Indus Water Treaty', *USI Journal*, January-March, 2010, pp. 75-87. Basing himself on declassified US State Department documents, the author says that the resolution of three disputes – Kashmir, Indus waters, and arms supply to Pakistan – formed the essentials of US policy towards South Asia during the Eisenhower administration (1953-61).
33. Ibid., p.81.
34. Cited in K. Warikoo, n.5, p.14.
35. Ibid., p.15.
36. Ibid., p.15.
37. 'One can view security as that which in language theory is called a speech act: it is not interesting as a sign referring to something more real – it is utterance itself in itself that is the act: by saying it, something is done.' Ole Waever, *Concepts of Security*, Copenhagen: University of Copenhagen Press, 1997, p. 221
38. United Jihad Council (UJC) is an umbrella organisation responsible for coordinating the activities in Pakistan of all jihadi groups.
39. Cited in *Final Settlement*, A Report by Strategic Foresight Group, at http://www.strategicforesight.com/finalsettlement/thesecret.pdf
40. Ibid.
41. Ibid.
42. Ibid.
43. Ibid. In 1990, Musharraf, then a Brigadier undergoing a one-year training at the Royal College of Defence Studies in London, wrote a dissertation titled, 'The Arms Race in the Indo-Pakistan Subcontinent: Conflicts with the Pressing Requirements of Socio-economic Development. What are its Causes and Implications? Is there a Remedy?' The dissertation

argued that Kashmir and water were interdependent, and that the rivers hold the key to any solution.

44. N.D. Gulhati, *Indus Water Treaty: An Exercise in International Mediation*, Bombay: Allied Publishers, 1973, pp. 440-447. In this article, David Lilienthal publicised his proposal of the Indus river basin development with external assistance.
45. Bashir A. Malik, *Indus Water Treaty in Retrospect*, Brite Books, Lahore, 2005, p.67
46. Ibid.
47. Ibid., p.161
48. Ibid., p.169
49. The full text of the Treaty is available on www.mea.gov.in.
50. See the Indus Water Treaty. Article II allocates all the waters of the Eastern Rivers for the 'unrestricted use of India', except as otherwise expressly provided in this article. The article provides for a transition period of ten years, beginning 1 April 1960, during which India would exercise some restraint as specified in Annexure-H on the withdrawal of waters from the Western Rivers, during which Pakistan would receive for unrestricted use the waters of the Eastern Rivers in accordance with this annexure. Article III states that, 'India shall be under obligation to let flow all the waters of the Western Rivers, and shall not permit any interference with these waters'. However, India is allowed certain uses of these waters for domestic use, non-consumptive use, and agricultural use as per Annexure-C, and the generation of hydroelectric power as set out in Annexure-D. Article IV contains a variety of provisions pertaining to both the Eastern Rivers and Western Rivers. The most important of these was about the construction of a system of works, which would 'accomplish the replacement, from the Western Rivers and other sources, of water supplies for irrigation canals in Pakistan which, on 15 August 1947, were dependent on water supplies from the Eastern Rivers'. Article VII is about 'Future Co-operation'. It states the 'optimum development of the rivers' by "mutual agreement to the fullest possible extent'. It relates to 'installing hydrological observation stations' and 'carrying out such new drainage works as may be required…' It also states '…the two parties may, by mutual agreement, co-operate in undertaking engineering works on the rivers'. However, if such engineering works 'affect the other party materially, it shall notify the other party of its plans and shall supply such data relating to the work as may be available…'
51. Speech by High Commissioner of India to Pakistan at the function organised by the Karachi Council on Foreign Relations and Pakistan-India Citizens Friendship Forum, on 3 April 2010, at http://www.india.org.pk/speech_high_commission.html.
52. Told to author in an informal conversation with a former Indian Indus Water Commissioner
53. Ibid., 'Out of the 33 projects, 14 are in operation, 13 are under construction, 2 are still at the proposal stage, 3 have been dropped or deferred and work on one project stands suspended.'
54. See Helmut R Kulz, 'Further Water Disputes Between India and Pakistan', *International and Comparative Law Quarterly*, 18(3), 1969, p. 727.
55. India had also compensated by releasing a similar amount of water as compensation from the Sutlej River to Pakistan. See, *Dawn*, 8 December 2008.
56. Pakistan lodged a complaint under Article IX (2) of the IWT and, on 15 January 2005, sent a request to the World Bank to appoint a neutral expert stating that a 'difference' had arisen. There were three complaints related to the project design; the size of the pondage; and the intake for the turbine, to which India did not agree. The World Bank

appointed Prof. Raymond Lafitte in May 2005 as the third party expert. Pakistan hired Peter Joseph Rae, a hydrologist from Canada. India hired Wolfgang Schwartz, a German consultant from Lahmeyer International. On the legal front, Fali Nariman argued for India, and Faisal Hussain Naqvi for Pakistan. On 12 February 2007, the NE gave the final determination. India had to show that the project was run-of-the-river, and did not store water in a reservoir. Pakistan had to prove that the project actually stored water in a reservoir behind gated spillway. See IDSA Task Force discussion with the Indus Water Commissioner. Also see, Arvind Gupta, 'Indus Water Treaty: Zardari ups the ante on Water Issues', IDSA Web Comments, 30 January 2009, at http://idsa.in/idsastrategiccomments/IndusWaterTreaty_AGupta _300109. Also see by the same author, 'Vicious anti-India propaganda in Pakistan on Water issues', IDSA Web Comments, 29 March 2010, at http://idsa.in/idsacomments/Viciousanti-IndiapropagandainPakistanonWaterissues_agupta_290310. Also see, Uttam Kumar Sinha, 'Water a pre-eminent political issue between India and Pakistan', *Strategic Analysis*, vol. 34, issue 4, 2010, pp. 482-485.

57. The non-paper, which in diplomatic parlance is an informal document exchanged for discussion, was handed over by Pakistan during the Foreign Secretary-level talks in February 2010. These notes were on Pakistan's wide-ranging concerns over reduction in the flow of the river due to climate change, for example, glacier melting, topographical conditions, pollution, etc. See Salman Bashir's, (Pakistan's Foreign Secretary) statement in *The Dawn*, 26 February 2010. He said Pakistan had informed the Indian side about the violation of the Indus Basin Water Treaty, the storage of water, India's plan to build more dams, the Kishangangahydel project, pollution in sources of water, and glacier melting. Also see, B.G. Verghese, 'Ideology Threatens Indus Treaty', *The South Asian Journal*, 25 March 2010.

58. '...there is an attempt to convert the technical issue to a political one', Satish Chandra, former Deputy National Security Advisor, Government of India, in conversation with the authors, 27 August 2010.

59. Asif Ali Zardari, 'Partnering With Pakistan', *Washington Post*, 28 January 2009.

60. Cited in Riyaz Punjabi, 'Indus Waters Treaty Human Security vs. Military Security', *Journal of Peace Studies*, vol.11 (4), October-December, 2004, at http://icpsnet.org/description.php?ID=332

61. Quoted in Syed Nazir Gilani, 'Water and Kashmir', *The News*, 11 May 2010, at http://www.thenews.com.pk/daily_detail.asp?id=238564

62. Saifuddin Soz's interview with the Journal of Peace Studies, 'Managing India's Water Resources', *Journal of Peace Studies*, vol. 13, Issue 04, October-December, 2006, at http://icpsnet.org/description.php?ID=429

63. See for details, F. A. Shaheen, M. H. Wani, S. A. Wani, and S. A. Saraf, 'Sustaining energy and food security in trans-boundary river system: Case of Indus basin', at http://www.riversymposium.com/index.php?element= SHAHEEN

64. Riyaz Punjabi, n.60. Cited in Utam Kumar Sinha, et al, 'Will the Indus Water Treaty Survive?', *Strategic Analysis*, vol.36, no.5, Sept-Oct, 2012, p. 745.

65. K. Warikoo, 'Indus Waters Treaty: View From Kashmir', 01 June 2006, at http://www.jammu-kashmir.com/insights/insight20060601a.html

66. Ibid.

67. Under Suhrawardy's leadership, Pakistan refused to accept the Bank proposals of 5 February 1954, and its Aide-Memoire of 21 May 1957. See, Bashir A. Malik, *Indus Water Treaty in Retrospect*, Lahore: Brite Books, 2005, p. 144-146.

68. There are conflicting assessments of the hydel potential. The J&K state puts it at 15,000 MW while the Centre for Monitoring Indian Economy (CMIE) estimates it at 7,487 MW. The restrictions placed on the use of water resources by the IWT have come in the way of the realisation of this potential. IWT gives a virtual veto to Pakistan to scuttle proposals for the harnessing of hydel potential on the western rivers.
69. G. D. Bakshi, 'Indus Water Treaty: Need for a Review', *Journal of the United Service Institution of India*, Vol. CXXXII, No. 548, April-June 2002, p. 223. Bakshi also argues: 'The Chenab is a river that flows for half of its length through the Indian territory. As such, natural justice demands that India have a much greater right on its water consumption than is permitted by the treaty', p.222
70. Zahir-ud-Din, 'Abrogate Indus Water Treaty?', *Rising Kashmir,* at http://www.risingkashmir.com/index2.php?option=com_content& task=review& id=18789
71. Lt. Gen. Eric A. Vas [Retd.], in 'Troubled Waters: Should We Revise the Indus Water Treaty?', at http://www.inpad.org/res67.html
72. Syed Nazir Gilani, n. 61.
73. The International Law Commission draft provisions, 'Responsibilities of States for Internationally Wrongful Acts', reproduced in the Yearbook of International Law Commission, 2001. Article 22 of Chapter V states that, 'Countermeasures in respect of an internationally wrongful act'.
74. S. Akhtar, "Emerging challenges to Indus waters treaty Issues of compliance and transboundary: Impacts of Indian hydro projects on the Western Rivers", at http://www.irs.org.pk/f310.pdf. Also see, A. Yaqoob, "Indus waters across 50 years: A comparative study of the management methodologies of India and Pakistan", at http://www.irs.org.pk/ecosocio/sps11.pdf
75. http://thesouthasianidea.wordpress.com/2010/04/03/war-or-peace-on-the-indus/
76. For the Text of the Indus Water Treaty, please see http://siteresources.worldbank. org/INTSOUTHASIA/Resources/223497-1105737253588/IndusWatersTreaty 1960.pdf

5

China and India: Hydropowers in South Asia

This chapter outlines China's upstream actions from three interconnected features: dams and diversions; the resultant hydropolitics and power asymmetry; and the impact of climate change on the glaciers of the Tibetan plateau. The chapter argues that it is essential for India to bring water issues into the core of the bilateral discussions with China, and to use all diplomatic tools to push towards a structured water dialogue that allows for lower riparian apprehensions and fears to be recognised and discussed. Equally important will be the need for both countries to study carefully the projections of future trends in water availability, flow patterns, and changes in climatic variables. Though certainly not easy, water as an area for mutual cooperation potentially opens up new possibilities based on sharing the benefits of the flow rather than merely determining and dividing the volume. The 'water rationality' view or 'water as a unifier' view would require a greater emphasis in hydro diplomacy, in exploring alternative institutional arrangements, and in effective dispute settlement mechanisms. The chapter also develops scenarios based on the understanding of current events and trends. The purpose, through the scenarios, is to clearly demonstrate that trans-boundary rivers have a high strategic content, and are crucial to peace and stability in the region. The management of such a vital resource does not operate in a vacuum but rather in a complex political and economic framework.

When water is scare, the importance of access to water becomes critical.

Driven by its increase in value and as a non-substitute resource, water potentially can become so precious that states will go the distance to possess it as well as become unwilling to share it. These tensions and stresses as well as possible conflicts can lead to altered relations in water basins. The relationship between riparian states can then be influenced by power, and the consequent power play. These power relations can lead to hegemony among the users of the same water. "Hegemony can be regarded as supported by authoritative leadership/authority."[1] Hegemony in a basin is determined by the behaviour of the hegemon which is based on enforcing the power over weaker riparian states in the river basin.

In 2006, Zeitoun and Warner developed an analytical framework which provides insight into the water hegemony between riparian states in an international river basin. This framework is based on the analysis of power relations within rivers. It indicates that a water-hegemon, which aims to consolidate control, uses different strategies, tactics, and power resources to achieve this control. When consolidated control is achieved, the water-hegemon will have the power over the whole basin. In the basin, there may also be attempts at counter-hegemony in order to change the power equation. China offers an interesting study as a supreme upper riparian, and analyses of the Brahmaputra river which, along with the Ganga and Meghna, forms an important river basin (second only to the Amazon), verify that China is most powerfully pursing consolidated control and resource capture in the basin. In other words it seems to be following a unilateral approach to the trans-boundary rivers. However, in more recent times, there seems to be a shift to more openness and a willingness to discuss and talk on water issues as well as share hydrological data and information. This can be viewed in the larger changing dynamics of increased dialogues, partnerships, and investments in the region.

Riparian relations are shaped and developed by varied interpretations of the use of river water and of the differing claims.[2] Upper riparian nations essentially base their claims on 'absolute territorial sovereignty' – that is, the right to use rivers unilaterally, regardless of lower riparian concerns. The lower riparian, on the other hand, claim 'absolute territorial integrity' of rivers, stressing that upper riparian actions should not affect the water flowing downstream. The two claims are incompatible. There are, however, accepted legal norms of 'equitable utilisation', 'no-harm rule', and 'restricted sovereignty' that riparian states work through, and frame negotiations and treaties accordingly to overcome such differing positions.[3] But, more often than not these norms in state politics and power dynamics are rendered meaningless. It is almost a vague notion that nations are entitled to a 'reasonable share of water'.[4] Given that there is no legally binding international treaty on water sharing, riparian relations will largely be influenced by the prevailing political dynamics and strategic considerations.

China is a critical player in the hydro-politics of the region. Its hydrological position is one of complete upper riparian supremacy. In contrast, India – another key player in hydro-politics is simultaneously an upper, middle, and lower riparian. India's middle riparian position increases its dependency (and thus water insecurity) on the headwaters of the rivers such as the Indus, Sutlej, and Brahmaputra which originate in the Tibetan plateau. China is equally water insecure, as explained earlier; but its insecurity relates to the uneven distribution of waters within its territory. China's hydrological position gives it enormous latitude in shaping larger political equations with its riparian neighbours. India, on the other hand, given its middle riparian position and its longstanding commitment to bilateral river treaties, has to assiduously balance the anxiety and concerns of its lower riparians (Pakistan and Bangladesh) without compromising its own water requirements.

For the river basin states other than China, being water-dependent on external sources is a hydrological reality. This, and the prevailing politics, shapes fears and perceptions. For example, the Mekong lower riparian countries will remain suspicious of China's upstream hydroelectricity projects.[5] Pakistan (which is heavily dependent on the rivers flowing in from outside its boundary) sees India, and not China, as an upper riparian aggressor. This, of course, greatly relates to the grievances that Pakistan has over the Indus Waters Treaty with India. Similarly with Bangladesh – the lowest riparian in the Ganga-Brahmaputra-Meghna basin – water becomes a political and emotional driver. As an upper riparian player, China would like the water debate in Pakistan and Bangladesh to be directed and contested with India, *without* highlighting its own hydroelectricity plans either on the Indus or the Yarlung Tsangpo (Brahmaputra).

While China has no formal water sharing arrangements with its neighbours, India has several treaties to address water issues with its neighbours, such as the 1960 Indus Waters Treaty with Pakistan, and the 1996 Ganga Treaty with Bangladesh. Water treaties commit India to a dialogue-based water sharing approach, and diplomatically become an important part of its neighbourhood policy. China, in contrast, would tend to take a strategic view of the water it commands; and, given its hydrological position, factor water as a tool, leverage, and a bargaining instrument in framing its regional policies. A snapshot of the riparian dynamics in the Himalayan watershed suggests that while there is considerable lack of trust on water issues between states, there is greater possibility of drawing India – rather than China – into the regional water debates, breaking political deadlocks through sensible water sharing arrangements, and resource development.

India and China: Contrasting Riparians

The deeply political question of 'who gets how much water, how and why,'[6] influences the behaviour of the riparian. Given their hydrological position and dependence on the Himalayan rivers, China and India are critical players in the hydropolitics of the region. As stated earlier, India is simultaneously an upper, middle and lower riparian. China's hydrological position, on the other hand, is one of upper riparian advantage. India's middle riparian position, increases its dependency (and is thus water insecure) on the head waters of the rivers sources such as the Brahmaputra, Indus, and Sutlej, which originate in the Tibetan plateau. China is equally water insecure, but its insecurity relates to the disproportionate availability or uneven distribution of waters within its territory, the majority of which is in the south (Tibet Autonomous Region), with the north and west excessively water stressed. In terms of per capita water availability, China ranks among the world's lowest.

The territorial source of the Himalayan rivers makes China far more water secure than India. In fact, China is probably the world's most independent riparian country.[7] This hydrological position gives it enormous latitude in shaping larger political equations with its riparian neighbours. In fact, China was one of the three countries that did not approve of the 1997 UN Convention on the Law of the Non-Navigational Uses of International Waterways.[8] On the Mekong River basin, China is only a dialogue partner in the Mekong River Commission (MRC), which was formed by the concerns of four lower-riparian countries – Cambodia, Laos, Vietnam and Thailand – in 1995. Though China's non-binding participation in the Mekong basin has increased, it is unlikely that it will join MRC as an active member because it does not want to formally commit to any arrangement of water sharing.

For India, as a Himalayan basin-state, being water-dependent on China is a hydrological reality and shapes its fears and perceptions. Apprehensions, in fact, are widespread. For example, the Mekong lower riparian countries remain suspicious of China's upstream hydroelectricity projects. Likewise, Kazakhstan and Russia are concerned over China's diversion of the Irtysh and Ili rivers. In the case of Pakistan and Bangladesh, India is seen as an upper riparian aggressor.

While China has no formal water sharing arrangements with the lower riparian countries, India has several treaties to address water issues with its neighbours, like the 1960 Indus Water Treaty with Pakistan and the 1996 Ganga Treaty with Bangladesh. With Nepal and Bhutan, treaties have been signed to share the benefits of water. Transboundary water treaties commit India to a dialogue-based water sharing approach. China, however, takes a strategic view of water and given its hydrological position uses water as leverage and a bargaining instrument in framing its regional policies. A snapshot of the riparian dynamics in the Himalayan watershed suggests that while there is considerable lack of trust on water issues between states, there is greater possibility of drawing India into a regional water debate and breaking political deadlocks through sensible water-sharing arrangements and resource development than is likely with China. The future water challenge in the Himalaya watershed is to draw China into a water dialogue.

India and China: Cooperation or Conflict

Water has emerged as a contentious issue between India and China, with complex inter-linkages among the social, environmental, economic, and political dimensions of the resource. As riparian neighbours and the two most populous countries, the waterscape is characterised by familiar challenges in the planning, design, and management of water resources in terms of quantitative, qualitative, and uneven distribution. Global warming impacts, rainfall pattern shifts, and expanding demands have also combined to put further pressure on water. There are worrying signs that the growing water shortages in the two countries, which rank amongst developing states with the lowest per capita reserve base, present the largest threat to food security.[9] With population increase and corresponding consumption patterns, it is projected that by 2030 the demand for water will be 50 percent higher in India and China.[10]

The leadership in both the countries has, from time to time, acknowledged the water problem as an existential threat. Back in 1998, Deputy Prime Minister Wen Jiabao expressed concern that the 'very survival of the Chinese nation' is threatened by the looming water shortage.[11] In his first Independence Day address in 2004, Prime Minister Manmohan Singh highlighted the issue of water, and raised it as one of the *saat sutras* (seven sectors) needing attention.[12]

At the same time, the political significance of water between India and China becomes crucial not only because of the supply-demand imbalance but also because the two countries share some significant glacial-fed rivers that originate from the Tibetan plateau. These even include the Indus and Sutlej on the western side. The Brahmaputra, on the eastern side of the plateau, is a precipitation-based river. Of the nine major tributaries of the Ganga that flow in from Nepal, the three principal tributaries – the Karnali, Gandaki, and Kosi – rise in Tibet.

Among these rivers the Brahmaputra, known as the Yarlung Tsangpo in Tibet, has become a source of anxiety for India. As a lower riparian, India's concerns revolve around future plans of water diversion as well as a series of dam projects undertaken by China on the Yarlung. Both will lead to flow fluctuations, and impact the local economy and ecology. Moreover, the Yarlung/Brahmaputra is intimately tied up with the issue of territory as well as China's claim on Arunachal Pradesh where the river enters India.[13] As scientific evidence mounts, worries also arise over the lack of shared information on the hydrological alterations due to the impact of global warming on the Himalayan/Tibetan glaciers.

While lower riparian angst and apprehensions beleaguer India, China's promotion of large-scale, capital-intensive water projects – with classic slogans like 'big diversions, big irrigation' – continues. Going by the 2011-2015 energy sector blueprint released in January 2013, far from restraining itself, Beijing plans to construct more hydroelectricity dams on the Nu (Salween), Lancang (Mekong), and the Yarlung river basins.[14] These are not only internationally-shared rivers but are also in ecologically and seismically sensitive areas. The blueprint is a reassertion of an aggressive 'supply-side hydraulic' approach of increasing storage capacity by building dams and reservoirs, water transfers, as well as prospecting and extracting groundwater.

This is the result of a combination of factors that includes food and energy needs, plans to meet the low carbon-intensity goals of the 12th Five-Year Plan (2011-2015), and the intensive lobbying of dam builders and electricity companies. By 2015, it is expected that some 120 gigawatts of hydroelectricity projects will be installed nationwide.[15]

Dams and Diversions

No region with shared international waters is exempt from water-related controversies and disputes. China's water needs and India's concerns will be a recurring theme in the relationship between the two countries. River uses are deeply subjective in terms of where, what, and how they are being used. With no legally binding treaty – except norms and principles as expressed in the 1966 Helsinki Rules and the 1997 UN Convention on the Law of the Non-Navigational Uses of International Watercourses – rivers are an unruly resource. Further, projections of a looming water crisis both in India and China raise questions about availability and distribution as well as legal difficulties. It is now a common national security refrain that a stable supply of water is paramount to a country's political, social, and economic stability. From a *realpolitik* prism, the preciousness of water often translates into possessiveness and, at times, resource aggressiveness.

China is an extremely thirsty country and is among the driest nations. The uneven distribution of water has been a critical stimulant to China's diversion plans.[16] Given such disparity, the idea of *Shou-tian* (or 'reverse flow') of the Tibetan rivers was proposed in 1988.[17]

Water projects have been part of the popular political consciousness since the foundation of the People's Republic of China (PRC) in 1949 led by Mao. Chinese engineers, who are a leading voice in decision-making, firmly believe that the diversion of rivers into the water scarce north and western region is crucial for growth and stability. The diversion on the Yarlung Tsangpo is part of the South-North Water Transfer Project (SNWTP),[18] which has yet to start. There are questions related to the technical, economic, and seismic feasibility of the project, but it has not been shelved. No doubt China will maintain diplomatic silence on its water diversion plans. Thus, this becomes an important threat multiplier, and helps create lower riparian fears.

China's growing requirement for energy is increasingly bringing water into the development process. The 12th five-year plan has made provisions for bases for the coal-industry across western China. These are seen as the new Strategic Emerging Industries, and will include coal mining, petrochemical and power plants.[19] Being water intensive, these industrial activities will require an enormous amount of water. The SNWTP diversion of the Yarlung and the other rivers (like the Lancang and the Nu) will probably get a reboot as the west-to-east coal-based industries develop. Given China's uneven water distribution, its energy needs, as well as its food requirements, it is difficult for Chinese planners to not consider water diversion as an option.

The upstream diversion of water is hugely scary for lower riparian states, and is quickly associated with evil intentions and deliberate acts of running a downriver state dry. India has been deeply concerned over the planned diversions of the Yarlung. Since this are yet to fructify – and increasingly seems difficult to achieve – its concerns seem to have diminished somewhat. The more immediate concern is regarding the major dams and storage projects that have already begun and are also being planned on the Yarlung, in order to 'push forward vigorously the hydropower base construction.'[20] China has proposed the construction of three dams on the middle reaches of the Yarlung: a 640 MW dam in Dagu; a 320 MW dam at Jiacha; and one at Jiexu, the capacity of which is not known. These three, along with the 510 MW dam in Zangmu which began in 2010, are part of a series of damming projects on the Yarlung.

For the Chinese who are used to water projects on a gigantic scale, the planned dams capacity are 'small'; but from the Indian point of view, these projects are sufficiently large to be storage dams, especially if the purpose is flood control and irrigation, as is the Zangmu.[21] Run-of-river (ROR) projects, as the Chinese planners officially describe them, can be misleading. The basic principle of the ROR dams is to return the waters to the river after it passes through the turbines. But what if they are not returned? A mechanism to 'trust and verify' is necessary.

Given the political equation between the two countries, China will use its riparian advantage as a response to the political temperature. It suits Beijing to be ambiguous, and not show enthusiasm towards formal arrangements on sharing design-related and hydrological information. Moreover, Chinese hydrologists explain that 'the Brahmaputra has plenty of water; it won't make any difference to India.'[22] To allay any unnecessary fears, even the Indian Water Resource Ministry has now openly stated that the Yarlung enters *India (as the Siang in Arunachal) with 78 bcm of water, and then collects 629 bcm before entering the Bangladesh border.*[23] On the question of Yarlung diversion (if it ever comes about), the Central Water Commission has suggested 'a 50 percent reduction of the 31.25 bcm currently available in the non-monsoon season and a reduction of 50 percent in power generation in the Upper Siang project.'[24] The figures suggest that India's concerns are not so much about water scarcity as it is about flood water release in the monsoon. *The solutions for Indian planners are essentially two-fold: build storage dams at various locations, and put effective flood mitigation programmes in place.*

India and China: Hydro Politics

Water relations can never be permanently settled since flows in river are not constant. The flows in turn are determined by seasonal variations and usage, particularly those that are non-consumptive in nature. Also, interventions and diversions on rivers impact flow. Political relations can easily be impacted by the changing quantitative and qualitative nature of rivers. Varied interpretations on the use of river waters have resulted in claims and counter-claims. However, more often than not, these norms in state politics and power equations are difficult to achieve. This also holds true for a nation's entitlement to a 'reasonable share of water.'[25]

Both India and China have a deep civilisational understanding of water, the shared common boundary, and the gusty rivers. The bilateral relationship is marked by inconsistency. At one level, trade between the two is promising and pushing forward in spite of odds; at another level, the boundary issue remains vexing and irksome.[26] Similarly, India's position on the Dalai Lama is a longstanding annoyance for China; for India, Chinese strategic reach in the Subcontinent is worrisome. At the international level, while the two converge as BASIC countries (Brazil, South Africa, India and China), in the climate change negotiations as well as in various other forums they have divergent views. India and China simultaneously cooperate, contest, and compete. Water relations have to be viewed and understood from this triangular perspective.

As economic powers, both are also 'planetary powers', consuming resources at a high rate with a heavy ecological footprint. Water is central to development and growth. The need for water in the food-energy nexus as well as for safe drinking will continuously challenge development goals.

With their growing economies, expanding ecological footprints, and rising political influence, China and India will need to be a part of the solution for a sustainable world economy. As two critically important riparian countries in Asia, it is an irony that there is an absence of institutionalised water cooperation. More so, with a climate future projected to be hotter and drier, and directly impacting water resources.

Thinking about water critically does not mean bellowing about water wars. History is also on the side of water cooperation. The record shows that water disputes do get resolved, even amongst bitter enemies and even as conflicts drag out over other issues. A good case in point is the 1960 Indus Waters Treaty between India and Pakistan, and the historic water treaty between Israel and Jordan in 1994 in which water was seen as a positive-sum game. India and China hydro-relations are evolving, and thus offer considerable scope for framing water as a resource for cooperation and exchange for joint work at the scientific and societal levels. However, peacebuilding and cooperative returns can best be achieved by assessing the vulnerability and strength of the hydrology in which China and India co-exist.

For India, hydro diplomacy will be a vital component of its neighbourhood policy which, from an expanded hydrological point of view, cannot exclude China. It is not easy to ignore the competitive nature of water between the two, as well as the significance of the Himalayan watershed from where the shared rivers originate as the hydrological faultline.

Climate Change and Himalayan Glaciology

In the coming years, the effects of climate change will take greater precedence over any physical changes associated with climate change. First, the food-energy-water nexus will be subject to various stresses and strains. Second, perceptions of a rapidly changing ecosystem will prompt nations to take several actions, many of them unilateral, to secure resources and territorial sovereignty.

Rivers carve the length and breadth of the Himalaya mountain system, physically linking upstream and downstream users. The glaciers, the dazzling source of these rivers, are vulnerable to various exogenous impacts, including global warming. Planning any water resource utilisation policy will have to take into account the assessment of the impact of climate change in terms of seasonal flow and extreme events.

The middle-latitude and high-altitude Himalayan glaciers contain one of the largest reservoirs of snow and ice outside the Polar Regions. This area is now commonly referred to as the 'Third Pole'. The contribution of snow and glacial melt varies from the eastern part of the Himalayas to the western. While the glaciers in the Karakoram region of north-western Himalayas are mostly stagnating, those in the western, central, and eastern Himalayas are mostly retreating. Of these, the western glaciers have shown the highest rate of retreat.[27] China's mountain glacier systems include the Himalayas, Karakorum, Tien Shan, and Altay mountain ranges, covering an area of about 59,425 sq. kms. The glaciers in India are located in the Himalayas, and cover about 8,500 sq. km.[28]

Scientific observations point to the possibility that 90 percent of the Himalayan glacier melt is directly caused by black carbon soot and other industrial processes.[29] Other studies point to the presence of debris, such as pebbles and rocks, as an additional factor. Simulations for the 2030s indicate all round warming over the Himalayan region by 1.7 to 2.2 degree centigrade, and the projected precipitation likely to increase by 5-13 percent.[30] The impact will gradually shrink glaciers, resulting in a decrease of water runoff in the long-term. In the short-term, the earlier water runoff from glaciers when combined with seasonal rains could result in flood conditions. There are 662 glaciers contributing to the Yarlung/Brahmaputra which, when compared to the Indus Basin (3583) and the Ganga Basin (1020), is considerably less, and translates to about 10-20 percent of glacial-fed contribution.[31] In the dry season, this is enormously valuable. Many recent studies on the overall glacier retreat and additional melt focus on dammed water or 'glacier lakes' (GLOF) that has the potential of generating dangerous outbursts of flooding.

Towards Water Dialogue

China perplexes the world. Its continued economic growth stirs the academic and research community into asking whether it will be the next superpower, and how the political landscape will change with its ascendancy. China's rise, whether peaceful or not, is also the subject of considerable debate. On trans-boundary water issues, the question of whether China can be a constructive upper riparian is crucial for a number of lower basin states, including India. The views expressed range from alarmist, fearful, and circumspect. There is also a quiet acceptance of China as the hydrological *supremo* as well as the reality of building political relations first and foremost to quell any hydro-aggression. All in all, these arguments make for a fascinating debate on China's hydro behaviour.

Before any formal water cooperation can be achieved, an understanding of water issues has to be developed. This requires diplomacy that is holistic, bold, and imaginative. Mechanisms for water cooperation have already been established between India and China and, for the time being, it is unrealistic to expect a treaty from China. In 2002, India entered into a Memorandum of Understanding (MoU) with China on the provision of hydrological information on the Yaluzangbu/Brahmaputra in the flood season. This information related to water level, discharge, and rainfall at three specified stations – Nugesha, Yangcun, and Nuxia – from 1 June to 15 October every year, which was utilised in the formulation of flood forecasts by the Central Water Commission. This particular understanding ended in 2007. A new MoU, with the same provisions and with a validity of 5 years, was signed in 2008. China provided hydrological data from the three stations during the monsoon of 2010.[32]

Another MoU was signed in April 2005 for the supply of hydrological information in respect of the Sutlej (Langquin Zangbu) in the flood season for a period of five years. This was renewed in 2010.[33] Clearly, China does not want a permanent mechanism on water issues. By reviewing and renewing the MoU, China dictates the proceedings as an upper riparian. During the visit of President Hu Jintao to India in November 2006, it was agreed to set up an Expert-Level Mechanism to discuss wider cooperation beyond flood season hydrological data to emergency management. Subsequently, a Joint Expert Level Mechanism (ELM) was constituted at the Joint Secretary level.[34] The ELM meets once a year, alternately in Beijing and New Delhi, and essentially focuses on the exchange of hydrological information and the smooth transmission of flood season hydrological data. It is very selective and limited, but forms the base on which future water cooperation can be further developed.

During the visit of Chinese Premier Li Keqiang to India in May 2013, serious time and discussion was given to water issues. However, India's proposal of a joint mechanism for better transparency on the dams being constructed on the Yarlung was diplomatically appreciated but failed to elicit a clear commitment from China. The Chinese followed the tested 2002 MoU format, and renewed the pact on sharing flood data during the monsoon for five years. The MoU says, 'China will provide to India twice a day the hydrological data of the Brahmaputra River in the flood season between 1 June–15 October'.[35] A forward step was taken by signing a new MoU for cooperation in 'ensuring water-efficient irrigation'.[36] When Indian Prime Minister Manmohan Singh visited Beijing in October 2013, the two countries reached further understanding to strengthen water cooperation on provisions of flood-season hydrological data and emergency management. It was also agreed that the flow of information provided by China would now commence from May 15 instead of June 1, starting 2014.[37]

However, these MoUs – often described as 'goodwill data' – are subject to periodic review. This is a reflection that China can ignore the water arrangement in times of heightened political tensions. Despite not being the most robust mechanism and, given China's non-committal position on trans-boundary rivers, this is probably the best that India could get. At one level India needs to trust China; but it also needs to monitor and verify the hydrological data – especially now that China is firmly going ahead with its plan to build a series of dams on the Yarlung/Brahmaputra.

The mechanism apart, any hydro-relations with China cannot ignore the power asymmetry in the basin. Currently, it can be argued that China is effectively mixing 'cohesion and compliance' with 'attraction and intimidation' – that is, what Antonio Gramsci termed 'a mix of force and consent.'[38] As China aspires to be a global leader, it will also have to adjust, and earn the respect of its neighbouring riparian. One way forward is to play a leadership role and develop a rules-based system that will help build an image of a non-threatening partner and careful listener.[39]

As a lower riparian to China and as an upper riparian to Pakistan and Bangladesh, India needs to express concern over China's upstream utilisation of water, and make it a core issue of bilateral talks. Raising concerns is perfectly acceptable in state relations. A typical recourse is to advocate the principles of international water law, however non-binding they may be. Raising concerns through such norms alerts and sensitises the international community, in spite of the fact that issues of 'equitable utilisation' and 'limited sovereignty' are always difficult and uncomfortable to agree upon. The principle of 'information exchange, notification and consultation' is crucially important in dealing with China, given the nature of the dams and water diversions. For example, withholding data on the flows of the rivers or on plans for building storage structures or dams or projects that divert water come under the 'no harm rules'.

India's own water sharing treaties with its lower riparians – Pakistan and Bangladesh – takes due account of the need for transparency. Under the Indus Waters Treaty, the Permanent Indus Commission meets regularly on river projects, and also undertakes site and field inspections. The Ganga Treaty with Bangladesh has provision for a Joint Committee, which examines difficulties arising over the implementation of the sharing arrangement. Of course India cannot ask China to reconsider signing the 1996 UN Convention on Non-Navigation Use of Water as India itself had abstained from voting in favour of it during the General Assembly debate, and has not yet ratified it. The 1996 Convention has come into force in August 2014. The Convention requires watercourse states to cooperate on the equitable and reasonable use and management of international watercourses, with a view to attaining their sustainable utilisation and adequate protection. At best, given China's position, the 1997 Convention offers only an enabling environment to improve dialogue and information exchange. It is, thus, important for India to continuously raise water issues with China, hoping gradually that this will mature into a water-sharing treaty.

India's riparian relations with China are exceptional and critical. India is multi-river dependent on the Brahmaputra in the east and the Indus and the Sutlej in the west. The Ganga, which originates in India, has nine tributaries joining it from Nepal, three of which (the Karnali, Gandaki and Kosi) arise in Tibet. Some figures indicate that about 354 BCM (billion cubic meters) of water flows into India from Tibet plateau.[40] If the goal of diplomacy is to turn potential water conflict into constructive engagement, then a water dialogue with China is necessary.

Turning the Equation

It is critical for India to articulate its middle riparian position, first to change the perception in the neighbourhood that India is a 'water hegemon'. This is often expressed by Pakistan and Bangladesh, in spite the robustness of the water treaties with them. Second, it is vital for India to draw China into the South Asian water equation through a multi-lateral basin approach, thereby sensitising China to downstream concerns and upstream responsibilities. Hydro-diplomacy has to be well nuanced and not always framed in legalistic terms. Rather, it should be put forward with a view to managing and engaging China.[41] This has significant political value when dealing with China over the Tibetan water resources. By raising the question

– however contested it might be – that China alone cannot be the stakeholder to the waters in Tibet, India creates the opportunity to articulate an ecological perspective and resource conservation principles.

By terming water resources in Tibet as a 'commons', India can draw international attention. This could possibly prompt China into a water dialogue with the downstream countries regarding ways of preserving and sharing the benefits of the Tibetan waters.[42] It needs to be remembered that China has a strong environmental constituency, with activists, scientists and journalists who, despite the odds, are sensitising local people and authorities on ecological issues and principles. In his 2007 report to the 17th National Congress of the Communist Party of China (CPC), President Hu Jintao called for more 'scientific development' on water issues including 'securing more clean drinking water, improving water conservation, water pollution prevention, restricting excessive water resources exploitation, and cutting water waste.'[43]

Tibet's ecology has been a key issue for civil society and powerful environmental groups – for example, the International Union for Conservation of Nature (IUCN) are campaigning for Tibet as a vulnerable area to be protected from rampant resource exploitation. In fact, in 2003, 7.1 million hectares in the Yunnan province, where the upper reaches of the Yangtze, Lancang, and Nujiang run parallel, was declared a World Heritage in 2003. Interestingly, China has ratified the Convention on the Protection of World Cultural and Natural Heritage which was adopted by UNESCO in 1972.[44] In 2004, Chinese NGOs opposed development projects on the Nujiang, prompting Wen Jiabao to take a difficult decision to halt the project pending a comprehensive environmental assessment. Likewise, environmentalists campaigned in 2008 to preserve the Tiger Leaping Gorge from the impact of a proposed dam. The government had to respond by moving the site of the dam away from the gorge.[45]

Warming and glacial melting has prompted unprecedented global pressure and action in areas where vulnerability is high. Tibet draws particular attention. It is becoming increasingly clear that rivers have more ecological functions than just providing water. The interaction of rivers and flood plains is one such vital function. With its weather patterns and hydrology systems, glacial conditions, and forest and soil functions, Tibet has an essential influence over Asia, providing sustenance to some of the world's most productive agricultural zones. No one expects Tibet to become a protected area; but it is in China's interest as well as in the interest of other riparian states to factor the ecosystem into Chinese water schemes.[46] It is also vital to involve local communities who integrate ecological values into everyday life, and have for long worried over Beijing's blatant mismanagement of the environment.

Responding to the science of climate change and global warming, China's environmental activists, scientists, and journalists have become a significant constituency, sensitising the local people, Chinese authorities, and the world at large on Tibet's diverse cultural and ecological wealth. This constituency could become an effective lobby, helping China's leaders steer the country towards sustainable development and better management of water resources. An environmentally conscious regime in Beijing, concerned with the impact of global warming, would be more open to conciliatory approaches to hydrological conflicts and would, perhaps, be more accepting of broad-based basin management of the Tibetan rivers.

Another important element of hydro-diplomacy would be for India to initiate a lower riparian coalition, stretching from the Ganga-Brahmaputra-Meghna basin to the Mekong, in order to draw China into a water dialogue. India's hydro-diplomacy has to ensure that the coalition is not seen as a counter-force or a challenger, or even a pressure group, but rather a concerned group seeking the opening of channels of communication and transparency with China on upstream usage based on the principles of 'equity' and 'no-harm'.[47]

To redefine a vital resource like water as a 'commons' should be an important part of the water dialogue between India and China in spite of the political sensitivity to such an approach in Beijing. As noted earlier, it is in everyone's long term interest, including China's, that Tibet's water resources are monitored as a sensitive ecosystem. India and China can together evolve a scientific community of glaciologists, hydrologists, seismologists, and climatologists studying and observing, collecting and comparing data, and building a knowledge base that helps harmonise development with nature. Interestingly, India and China have become observers in the Arctic Council, and their endeavours to observe and share notes on the geo-physical changes in the Arctic could be a positive experience for the study of the glaciers of Tibet.

Hydrological Scenarios: the Shape of things in 2030

Scenario 1: Tibet, from autonomy to complete annexation. Tibetan waters for China and China only

A supremely confident China, with the second largest economy and a formidable military power, is taking centre stage in Asian politics. While being military strong and economically dynamic, it is equally resource hungry. Its thirst for oil makes it the largest importer of natural resources which it has, over the last three decades (since 2000), made viable by secure political linkages with African and Latin American countries. Its accessibility to the warm waters of the Arabian Sea through the Gwadar Port near Karachi and the development of the Karakoram Highway gives it enormous flexibility. Its infrastructural aid and assistance over the last few decades has helped it to build unmatched political and diplomatic connections with Islamabad and Yangon (Myanmar), irrespective of the type of regimes in these countries. India, its strongest challenger, has been 'encircled' through Pakistan in the West and Myanmar in the East.

China continues to be thirsty for water as well, and considers water an unquestionable national priority. Tibet no longer remains an autonomous region, and its complete annexation has been prompted by securing water resources by completing many of the diversion plans initiated in the first decade of the 21st century, and many more are in the operational stages. China's plan of rapid economic development and encouraging the influx of Han Chinese into Tibet (since 1990) has slowly but surely paid off. A new generation of Tibetans, aware of China's awesome dominance, is less inclined to struggle for freedom, and has now learnt to coexist with the demographic changes. The resources of Tibet are now in full control and for the sole interest of China. The legacy of Hu Jintao, the paramount leader (2002-2012), and his mega projects continues to inspire the current leadership of Xi Jinping, still dominated by hydrologists and engineers, to build more transfer-of-water-projects. Way back in 1952, Chairman Mao Zedong had innocently remarked, 'the south has a lot of water, the north little...If possible, it is okay to lend a little water.' In fact, Beijing's total control over Tibet in effect is the 'total' control of its water resources. 'Absolute territorial sovereignty' and 'hydro-egoism' dictate water relation with its neighbours. The following classic lines *have become a reality for the lower riparians.*

> *He who holds Tibet dominates the Himalayan piedmont; he who dominates the Himalayan piedmont, threatens the Indian subcontinent; and he who threatens the Indian subcontinent may well have all of South-East Asia within his reach, and all of Asia.*[48]

This reality has forced the lower riparians to live-and-let-live, with a China unwilling to concede on water sharing and look within to improve water-use efficiency of irrigation projects and surface water systems.

Scenario 2: Tibet, from autonomy to partial independence. Tibetan water embraces basin based 'conjunctive use' than single-country 'water management'

Chinese lopsided development policies and unrestrained exploitation of natural resources has led to a shaky China, crippled with mounting domestic challenges from income disparities, social pressures, corruption, and environmental devastation. All this is impeding its economic progress. The Communist Party is troubled, and hard pressed to retain the public acceptance of the party's dominant role. Chinese leaders are finding ways to bring in greater political pluralism and accountability in order to restore public confidence.

The difficult situation is further compounded by heightened nationalism in Tibet where the radicals, after the death of the Dalai Lama, are pitching for complete freedom. The US Congress now officially encourage campaigns for independence. Beijing is in a bind. It has long feared that any concessions towards Tibet, which it has had to concede gradually, would result in more uncomfortable demands. With the centre not so strong and the periphery revolting, its worst fears have come alive. The growing anti-dam lobby has pushed the Chinese leadership to cancel many of its projects. Following on the lines of Prime Minister Wen Jiabao who cancelled 13 dams project on the Nujiang in 2004, the saner voices in the Chinese leadership are in favour of sustainable development and a wider consideration for social and environmental impacts. Moreover, these have grown in number, and are slowly marginalising the dam lobby. Without complete independence but with full control of its domestic affairs, the Tibetans have an equal participation in the management of the resources (not just needs-based but rights-based development), with Mainland China responsible for defence and foreign affairs. The decontrol of water resources in Tibet has ushered in a new framework on 'water resources management' and 'hydrosolidarity' that includes multi-purpose beneficial utilisation of the resource, with active participation of the basin states of South Asia.[49]

Scenario 3: Ecological concerns put the spotlight on Tibet's glaciers and water resources. China responds positively

The science of climate change on the subject of warming and glacial melting is nearer the truth as never before. This has prompted unprecedented global pressure and action in areas where vulnerability is high. Tibet draws particular attention. The International Union for the Conservation of Nature (IUCN) – the largest and the most powerful environmental pressure group along with the UN – declares the entire province of Tibet to be a protected area.[50] China's initial resistance, based on suspicion and interference, gives way to a sensitive understanding of the ecosystem. This has widespread approval of the local Tibetans who consider ecology as a way of life, and have for long derided Chinese blatant mismanagement of the environment. This refreshing development has come about from within China. Responding to the science of climate change and global warming, Chinese environmental activists, scientists, and journalists have, despite all odds, worked hard to sensitise the local people, the authorities, and the world on Tibet's diverse cultural and ecological wealth. This growing and concerned constituency has become an effective lobby, helping China's leaders steer the country towards sustainable development and better management of water resources. Domestic legal mechanisms are strengthened with SEPA (State Environmental Protection Agency) ensuring that the highest standards are applied on meeting environmental impact assessment law (EIAL). Concerned with the impact of global warming, an environmentally conscious regime in Beijing makes conciliatory moves, allowing for broad-based basin management of the Tibetan rivers as well as framing effective bilateral river treaties.

Scenario 4: A 'coalition of lower riparians' challenges China on water issues

Probably nothing thrills China more than water, but also, nothing bereaves it more. A strong and confident China displays 'hydro-egoism', and a riddled and tentative China reconciles to 'hydro-solidarity'. Irrespective of China's rise, the thirst of the lower riparians have resulted in a cohesive South and Southeast Asian grouping, built on lower riparian concerns and based on the principles of 'equity' and 'no-harm'. This grouping also challenges China's unrestricted exploitation of the water resources in Tibet.[51]

This has expanded in recent times to include Kazakhstan and Russia who, as downstream countries, are grieved over China's diversions on the Irtysh and Ili rivers. There is much concern in Cambodia, Laos, Thailand, and Vietnam over a large number of dams being built on the Lancangjiang River since 2020, which has considerably reduced the flow of the Mekong River downstream, affecting agriculture, fishing and navigation. Myanmar is equally disturbed over the Salween River as the influential dam lobby in China has won the day, and restarted building the 13 dams on the Nujiang River.

Taking this as a big opportunity, India unshackles it middle riparian constraint, which had hindered its riparian equation with its neighbours for many years, and puts the spotlight on China as the actual riparian aggressor. China is undoubtedly perturbed, and views India's role with great suspicion and nervousness. India has an important pole in the multipolar international system, serving as a political and cultural bridge between a rising China and the West. Moreover, it is also driving towards building partnerships with many countries which it calls its 'arc of interest'. These partnerships are based on shared concerns, particularly on shared watercourses. Within its neighbourhood, India is now viewed with less suspicion. Building a riparian coalition has helped raise Tibet's water resources – which are vital not only to the Indus and GBM Basin but also to the Salween and Mekong to becoming a regional concern, and ensuring that its natural flow is not reversed by China's artificial diversion plans. A need to redefine a vital resource like water as a 'commons', as well as preserving and sharing the waters of Tibet, has gathered momentum. While defining water as a 'commons' has been politically sensitive by clashing with national jurisdictions it has, nonetheless, gained international credibility and sympathy. Sustained political and environmental pressure on China enables Tibet's water resources to become a 'commons' under global protection, and monitored as a sensitive ecosystem.

Scenario 5: Pakistan-China become a riparian friend-in-need to counter India

Water relations between India and Pakistan continue to affect bilateral relations. As a lower riparian, Pakistan habitually raises the issue of water with India, but finds little international audience. However, it has found a new meaning to its relationship with China on hydro-cooperation. The two projects in Gilgit-Baltistan, the Diamer-Bhasa dam on the Indus, and the

Bhunji hydro-project on the confluence of the rivers Indus and Gilgit, have added a new strategic dimension to the relationship, and paved the way for other build-operate-transfer projects. In spite of the fact that China has built its own storages and dams on the headwaters of Indus in the Ngari Prefecture of Tibet[52] (thereby manipulating the flow of the Indus waters), Pakistan considers its dependency on China less threatening than its water dependency on India. For Pakistan, India remains the perennial threat, and is willing to compromise its riparian position vis-à-vis China to strategically counter balance India. The dam partnership between two of the world's biggest dam builders has led to serious ecological and societal concerns, and the Pakistani regime is under pressure to rethink its approach to big structures on the rivers. Bearing the brunt of its down riparian position, Sindh is up in arms with widespread civil unrest making Pakistan an unstable and high-risk state. Pakistan's dependency on China has come with a telling cost, and countering India in its overall game plan has not worked.

Scenario 6: India takes the burden out of the Indus Water Treaty, and evolves pragmatic approaches.

Having observed carefully the strategic alliance of China with Pakistan, its own water needs and requirements, and the different constituencies within India with different perspectives on the Indus Water Treaty, the government of India has taken a middle-path of exploiting the potential permissible within the treaty. The decision has not been easy, given the popular pressure to abrogate the Treaty as Pakistan continues to abet and aid terrorist activities directed at India. By 2030, India's position as an upper riparian is one of pragmatism, drawing international kudos. Prompted by China's own storage designs on the upper reaches of the Indus, India has already constructed storage capacity of 3.6 Maf on the western rivers that the Treaty allowed but which, for over 50 years, did not achieve. With this acquired capacity, India's stance on the Indus waters is stronger and firmer. India continues to 'talk' to Pakistan but not 'negotiate', and the talks centre on 'water needs' and not 'water rights'.

By 2030, it has also factored in the needs and aspirations of the people of Jammu and Kashmir, and completed the 330 MW Kishanganga hydropower project as well as the Tulbul Navigation Project. India's robust economy and stability gives it the capability to use the 'priority principle' of the Treaty, and complete other projects on the Indus basin. It has also taken into account the poor maintenance of the existing barrages that allowed for about 2-3 Maf of water flowing easily into Pakistan. Having not terminated the Treaty, and having mastered the use of water as a political leverage, and with Pakistan's inability to reign in terror, India has forced Pakistan into agreeing to the 'modification' of the Treaty, thus allowing it to develop and harness the potential of the Western Rivers. Pakistan is now compelled to buy peace for water from India.

The Way Forward

Benefit sharing that shifts away from a volume-driven approach is widely regarded as a rational solution to contentions over water. While this is attractive, it is certainly not easy to implement. For India to initiate a cooperative framework, benefit sharing must offer rewards greater than those of unilateral action. Identifying the benefits can itself be a contentious exercise, involving trade-offs and rigorous economic accounting. The immediate derivable benefits between India and China on the Yarlung/ Brahmaputra are the mitigation of floods and the potential for joint hydropower generation. These would add positively to the development activities of the two countries. A case in point is the Itaipu dam between Brazil and Paraguay. Despite tough negotiations, the two countries signed what is the world's largest and most successful hydro-energy cooperation in 2009.

The accuracy and regular availability of credible hydrological data are vital to the effectiveness of any trans-boundary water arrangement. India and China would need to collaborate on data generation and constantly upgrade data-sharing mechanisms based on the MoU signed in 2002. Such information will be instrumental in instituting early warning systems and better flood management in north-east India. This will also help in removing irritants over differences in data, which both sides use to justify their preferences.

Conclusion

China's upper riparian position and its enormous domestic requirements make water a critical resource which is fundamental to its development. Its upstream actions evoke different levels of concern. China's quest to dominate the flow will continue, with hydraulic manipulations and civil engineering interventions. The enthusiasts of such a paradigm, which Karl Wittfogel described as leading to a 'hydraulic society,'[53] have 'emptied water of its historical, cultural and ecological properties.'[54]

This chapter has emphasised that water relations between India and China have a very limited conflict potential though, as rising powers, there will be rising tensions over trans-boundary rivers. It cannot be denied that China's hydrological position gives it the strategic width to deal with India, a lower riparian. It argues that mechanisms (MoUs) are in place since 2002 on the Brahmaputra and the Sutlej; but they need to be continuously verified and monitored. Despite the enforcement of the 1997 UN Convention, this chapter's emphasis has been more on the effect of power relations as well as the exertion of power by the hegemon.

While China is showing greater sensitivity to India's lower riparian apprehensions than before, it has also clearly drawn the lines, and is unlikely to move beyond the 'expert-level mechanism' to a more comprehensive treaty. China is comfortable with *not* having water treaties, and prefers to engage bilaterally in limited formats. Transparency on hydrological data-sharing and a willingness to listen is part of its effort to build a positive image as a responsible upper riparian power. However, it is equally attentive to its own enormous water requirements and, to that effect, will be uncompromising. Externally, China will use its riparian advantage to define political equations with its neighbouring countries. It suits Beijing to be ambiguous on water projects, and there should be no doubt that it will not hesitate to employ water as a strategic tool and an important diplomatic leverage – or what has been expressed in the strategic circle as a 'non-confrontationist aggression', if it becomes necessary.

There is, however, space for a hydro-diplomacy that engages China in a broad-based water dialogue – one that aims at not only a bilateral but also a multilateral, basin-wide approach. The principle of 'information exchange, notification and consultation' and the principles of 'no harm' are sound points on which to consistently engage China. While Beijing will continue to enjoy its riparian position and determine relationships based on power, India's counter approach should be to take a bold stride towards creating awareness about water resources and the ecological significance of Tibet. The water dialogue between India and China should use science and research capacities to provide the knowledge backbone needed for water management.

NOTES

1. Mark Zeitoun and Jeroen Warner, 'Hydro-hegemony: a Framework for Analysis of Trans-boundary Water Conflicts', *Water Policy*, 8, 2006, pp. 435-460.
2. Riparian nations are those nations 'across which or along which a river flows, have legal rights to use the water of river'. Please see, http://www.waterencyclopedia.com/La-Mi/Law-International-Water.html
3. The rule of 'equitable utilisation' is based on the concept that an international drainage basin is a coherent legal and managerial unit, and embodies a theory of restricted sovereignty under which each nation recognises the right of all riparian nations to use water from a common source, and the obligation to manage their uses so as not to interfere unreasonably with like uses in other riparian nations.
4. While this is enshrined in the UN Convention on the Non-Navigational Uses of International Watercourses (1997), it is not binding because the convention is not in force as only sixteen (out of the required thirty-five) countries have ratified it. The UN General Assembly approved the convention by 104-3 in 1997. There is also the 'no-harm rule' in the convention, which requires riparian nations to take all 'appropriate measures' to prevent causing harm to other watercourse nations.
5. Likewise, Kazakhstan and Russia will be concerned over China's diversion of the Irtysh and Ili rivers.
6. See n. 1. Mark Zeitoun and Jeroen Warner, 'Hydro-hegemony: a Framework for Analysis of Trans-boundary Water Conflicts', *Water Policy*, 8, 2006, pp. 438.
7. Independent in terms of sources of rivers originating in its territory, and not being dependent on the sources of the headwaters from other countries.
8. The other two were Turkey and Burundi.
9. The per capita water availability, measured in cubic meter/person/year, on an average from 2003-2007 for China was 2138, and for India it was 1719. See 'Computation of Long-term Annual Renewable Water Resources by Country (in km^3/year average): China,' AQUASTAT, Food and Agriculture Organisation of the United Nations, 1 August 2013, at http://www.fao.org/nr/water/aquastat/data/wrs/readPdf.html?f=WRS_CHN_en.pdf
10. McKinsey and Company, *Charting our Water Future: Economic Frameworks to Inform Decision Making*, Munich: 2030 Water Resources Group, 2009, http://www.mckinsey.com/App_Media/Reports/Water/Charting_Our_Water_Future_Exec%20Summary_001.pdf.
11. "Drying Up," *The Economist*, 19 May 2005, at http://www.economist.com/node/4000643.
12. See 'Prime Minister's Independence Day Address,' Prime Minister's Office, 15 August 2004, at http://pmindia.nic.in/speech-details.php?nodeid=6.
13. China refers to Arunachal Pradesh as 'South Tibet' (*Zangnan*), and bases its claim on historical ties between the Lhasa and Tawang monasteries. These claims are frequently reiterated, giving the impression that for China, Arunachal is now also about acquiring the vast water resources there. See IDSA Task Force Report, *Water Security for India: The External Dynamics*, New Delhi: Institute for Defence Studies and Analyses, 2010, p.48, at http://www.idsa.in/sites/default/files/book_WaterSecurity.pdf.
14. Li Jing, 'Ban Lifted on Controversial Nu River Dam Projects,' *South China Morning Post*, 25 January 2013, at http://www.scmp.com/news/china/article/1135463/ban-lifted-controversial-nu-river-dam-projects.
15. Ibid.

16. This is the region south of the Yangtze River, which accounts for roughly 36 per cent of Chinese territory, and has 81 per cent of its water resources. The territories north of the Yangtze make up 64 per cent but have only 19 per cent of water resources.
17. The idea of 'reverse-flow' was the brainchild of hydrologist Guo Kai, who was intrigued by Chinese geologist Weng Winhao's theories on Tibetan hydrographic net. Thus, the idea of diverting Tibetan rivers was planned. Guo calculated that if the water from the Nujiang, Lancang, Jinsha, Yalong, and Dadu rivers in Tibet were diverted through the Aba divide, a solution could be provided for the crippling water shortage felt in north and northwest China. In 1988, Guo's plan caught the attention of the military. In 1989, a committee headed by top generals was formed to study the proposal. The first field survey of furthering projects on the 'Great Western Route' took place between 18 May and 22 June, 1999. Guo Kai was part of the expedition. It now has wide support in the PLA. In 2005, Li Ling's *Save China through Water from Tibet* [cited in Claude Arpi, "Himalayan Rivers: Geopolitics and Strategic Perspectives" in *Water: Culture, Politics and Management*, India International Centre, New Delhi, Pearson, 2010, p.10] further caught the attention of various ministries and planners. See IDSA Task Force Report, *Water Security for India: The External Dynamics*, New Delhi: Institute for Defence Studies and Analyses, 2010, p.48, at http://www.idsa.in/sites/default/files/book_WaterSecurity.pdf.
18. SNTWP has three routes of diversion: the central, the eastern, and the western. The first two routes are on the internal rivers. The western route is planned on the Yangtse River, and to divert about 200 billion cubic meters (bcm) of water annually.
19. APCO Worldwide, *China's 12th Five-Year Plan*, 10 December 2010, at http://apcoworldwide.com/content/PDFs/Chinas_12th_Five-Year_Plan.pdf.
20. Quoted in Ananth Krishnan, 'China Gives Go-ahead for Three New Brahmaputra Dams,' *The Hindu*, 30 January 2013, at http://www.thehindu.com/news/international/china-gives-goahead-for-three-new-brahmaputra-dams/article4358195.ece.
21. Himanshu Thakkar, 'Chinese Checkers,' *Hindustan Times*, 12 February 2013, at http://www.hindustantimes.com/editorial-views-on/ColumnsOthers/Chinese-checkers/Article1-1010765.aspx.
22. Zhou Wei, 'Divided Waters in India,' *Chinadialogue*, 20 September 2011, at http://www.chinadialogue.net/article/show/single/en/4539-Divided-waters-in-China.
23. Standing Committee on Water Resources (2009-2010), *Fourth Report*, August 2010, p. 43, at http://www.rgics.org/test/wp-content/uploads/2012/10/WORKING-OF-BRAHMAPUTRA-BOARD..pdf.
24. 'Parched Tiger to Thirsty Dragon: "No Concerns about Dams on Brahmaputra"', *Water Politics*, 8 February 2011, at http://www.waterpolitics. com/2011/02/08/parched-tiger-to-thirsty-dragon-no-concerns-about-dams-on-brahmaputra/.
25. This is enshrined in the UN Convention on the Non-Navigational Uses of International Watercourses (1997), which has now come into force on 17 August 2014 with Vietnam becoming the 35th instrument of ratification on May 21, 2014. The UNGA had approved the UN Convention by 104-3 in 1997, and was opened for signature from 21 May 1997 to 21 May 2000. States, however, continued to ratify, accept, approve, or accede to the Convention indefinitely. The 'no-harm rule' in the Convention, which states riparian nations to take all 'appropriate measures' to prevent causing harm to other watercourse nations. See United Nations, *Convention on the Law of the Non-navigational Uses of International Watercourses*, adopted by the General Assembly of the United Nations, 21

May 1997, at http://untreaty.un.org/ilc/texts/instruments/english/conventions/8_3_1997.pdf.

26. India and China's bilateral trade peaked in 2011 with USD 73.9 billion, but fell by 12 percent in 2012 to USD 66.47 billion. The expected target is to reach USD100 billion by 2015. See Ananth Krishnan, 'India's Trade with China Falls 12%,' *The Hindu* 10 January 2013, at http://www.thehindu.com/business/Economy/indias-trade-with-china-falls-12/article4295117.ece.
27. Study carried out by the Department of Geography, University of California, Santa Barbara.
28. H.C. Nainwal, 'Retreat of Himalayan Glaciers: Examples from Garhwal Himalaya,' at http://nidm.gov.in/idmc/Proceedings/LandSlide/A2-8,%20 Nainwal%20H%20C.pdf.
29. Recent studies indicate that the accumulation of black carbon is accelerating the shrinking of the Himalayan glaciers, and these are not only coming in from the East (China) but largely from South Asian countries. See Wu Wencong, 'Third Pole Glaciers Shrinking: Affected by Black Carbon,' *China Daily*, 25 March, 2013, at http://www.chinadaily.com.cn/china/2013-03/25/content_16341312.htm.
30. 'Climate Change and India: A 4x4 Assessment', Ministry of Environment and Forests, Government of India, New Delhi, November 2010, p.12, at http://www.moef.nic.in/downloads/public-information/fin-rpt-incca.pdf
31. 'An Overview of Glaciers, Glacier Retreat, and Subsequent Impacts in Nepal, India and China', WWF Nepal Report, p. 30, at http://www.wwf.or.jp/activities/lib/pdf_climate/environment/Overview_of_Glaciers.pdf
32. 'India-China Co-operation', Government of India, Ministry of Water Resources, New Delhi, at http://mowr.gov.in/index3.asp?sslid=372&subsublinkid= 290&langid=1.
33. Ibid.
34. Ibid.
35. Gargi Parsai, 'India, China Renew Flood Data Pact on Brahmaputra', *The Hindu*, 20 May 2013, at http://www.thehindu.com/news/national/india-china-renew-flood-data-pact-on-brahmaputra/article4732965.ece.
36. Ibid.
37. Uttam Sinha, *India-China Hydrorelations, China-India Brief* 15, Lee Kuan Yew School of Public Policy, 23 October –12 November 2013, at http://lkyspp.nus.edu.sg/cag/publication/china-india-brief/china-india-brief-15
38. Antonio Gramsci, *Selections from the Prison Notebooks of Antonio Gramsci*, (eds. and trans.) Quintin Hoare and Geoffrey Nowell Smith, New York: International Publishers, 1971. See also, Mark Zeitoun, Naho Mirumachi and Jeroen Warner, 'Transboundary Water Interaction II: The Influence of "Soft Power"', *International Environment Agreements: Politics, Law and Economics*, 11/2, 2011, p. 164.
39. The phrase is similar to the one expressed by David Shambaugh: '...most nations in the region now see China as a good neighbour, a constructive partner, a careful listener and a non-threatening partner.' See, David Shambaugh, 'China Engages Asia: Reshaping the Regional Order', *International Security*, 29/3, winter 2004-2005, p.64.
40. 'Water Security for India: The External Dynamics', IDSA Task Force Report, p.44, at http://www.indiaenvironmentportal.org.in/files/book_WaterSecurity. pdf
41. Sujit Dutta, 'Managing and Engaging Rising China: India's Evolving Posture', *Washington Quarterly*, 34/2, 2011, pp.127-144.
42. Uttam Kumar Sinha, 'Tibet's Watershed Challenge', *The Washington Post*, 14 July 2010,

at http://www.washingtonpost.com/wp-dyn/content/article/2010/06/13/AR2010061303331.html.

43. See text of the speech by Hu Jintao, 'Hold High the Great Banner of Socialism with Chinese Characteristics and Strive for New Victories in Building a Moderately Prosperous Society in All', *China.org.cn*, 25 October 2007, at http://www.china.org.cn/english/congress/229611.htm.
44. See United Nations Educational, Scientific and Cultural Organisations (UNESCO), "Three Parallel Rivers of Yunnan Protected Areas," at http://whc.unesco.org/en/list/1083.
45. 'Paradise Lost at Tiger Leaping Gorge', *China Digital Times*, 7 November 2008, at http://chinadigitaltimes.net/2008/11/paradise-lost-at-tiger-leaping-gorge/.
46. There is an example. The 7.1 million hectare in Yunnan Province, where the upper reaches of Asia's three great rivers (Yangtze, Mekong and Salween) run parallel from north to south, was declared a World Heritage in 2003. The Convention concerning the Protection of World Cultural and Natural Heritage was adopted by the General Conference of UNESCO on November 16, 1972. China ratified it. See UNESCO, 'Three Parallel Rivers of Yunnan Protected Areas', at http://whc.unesco.org/en/list/1083.
47. The community of co-riparian states is an important principle in the waters of an international river. According to Salman M.A. Salman, 'The basis is that the entire river basin is regarded as an economic unit, and the rights over the waters of the entire river are vested in the collective body of the riparian states, or divided among them either by agreement or on the basis of proportionality. Clearly, this is an ideal principle that overlooks sovereignty and nationalism.' Also, the 'Equality' and 'No harm' principles are based on 'limited territorial sovereignty' and 'limited territorial integrity' as opposed to 'absolute territorial sovereignty' and 'absolute territorial integrity'. See Salman M.A. Salman, 'Helsinki Rules, the UN Watercourses Convention and the Berlin Rules: Perspective on International Water Law', *Water Resources Development*, 23/4, December 2007, pp. 627-628.
48. See, George Ginsburg and Michiel Mathos, *Communist China and Tibet: The First Dozen Years*, Martinus Nijhoff, 1964.
49. The concept of 'water management' has historically meant the manipulation of water for specific uses and the development of water-based projects like dams, canals and other such structures. The concept of 'water resources management' regarded as 'rational' and 'integrated', embraces the 'conjunctive use' of "surface and underground water resources, the control of their quality and harmful effects, their sustainable development for multi-purpose beneficial utilisation and the administration of corresponding use rights." See, Bernard J Wohlwend, 'Equitable Utilisation and the Allocation of water Rights to Shared water Resources', at. http://www.bjwconsult.com/EQUITABLE.PDF. p. 2.
50. There is an example. The 7.1 million hectare in Yunnan Province, where the upper reaches of Asia's three great rivers (Yangtze, Mekong and Salween) run parallel from north to south, was declared a World Heritage in 2003. The Convention concerning the Protection of World Cultural and Natural Heritage was adopted by the General Conference of UNESCO on 16 November 1972. China ratified it. See, http://whc.unesco.org/en/list/1083
51. Please see, n. 42.
52. Alice Albinia, *Empires of the Indus*, London: John Murray, 2008. The author describes a dam being built on the upper reaches of the Indus in Tibet. Reports have confirmed that Senge-Tsangpo Hydropower station on the Indus has a capacity of 6,400 MW.

53. Karl Wittfogel was noted for his work titled *Oriental Despotism* first published in 1957. Wittfogel theorised that the control of water supply for irrigation was the basis of the Asiatic mode of production, and of a powerful, exploitative bureaucracy – a theory that came to be known as 'hydraulic monopoly'. See Karl A. Wittfogel, *Oriental Despotism: A Comparative Study of Total Power*, New Haven: Yale University Press, 1957.
54. Rohan D'Souza, 'From Damming Rivers to Linking Waters: Is this the Beginning of the End of Supply-Side Hydrology in India?' In Terje Tvedt, Graham Chapman and Roar Hagen (eds.) *A History of Water, Volume 3: Water, Geopolitics and the New World Order*, London: IB Tauris, 2011, p.357.

CONCLUSION
'LET THE RIVER FLOW...'

Let the blind man say
I can see again
Let the dead man say
I am born again
Let the river flow
Let the river flow

– Song by Darrell Evans, 1991

Water is acknowledged as the bloodstream of the biosphere. Water security is like 'gossamer' that links together food, energy, economic growth, and human security challenges. The complexities involved on water issues are multi-pronged in South Asia. Delineating the complexities involved on water issues is critical to regional policies and thus peace and stability. There are early warning signs that the region needs to be prepared for, and work together, to mitigate the dangers of water security. South Asia is an important region. It constitutes some one fifth of the world's population. The 21st century has been called the Asian Century and South Asia is an important sub-region of Asia. The region is a fertile ground for wide-ranging studies from inter and intra-state conflicts and nuclear proliferation to development, political economy and human security. The region also offers a hydrological perspective and the interlinked challenges.

The region is already witnessing, at a rapid rate, water wasting and overuse. A series of regional water 'bubbles' to support economic growth, especially in the agricultural sector, has left groundwater stocks seriously depleted. Clearly, the consequences for regional economic and political stability will be serious. Unlike energy, water has no substitutes or alternatives and, therefore, managing water will be essential to maintaining the economic web. Water is a systemic crisis and an urgent political issue that requires government engagement in its management and reform. The various dimensions of water – including the fact that it is the most shared resource – demonstrates that wide collaboration and cooperation, although difficult, is the only effective way to address it. It also offers an opportunity led by government, a multistakeholder, effort to improve the management of our future water needs by bringing together state institutions, business, and civil society.

Future water problems and their solutions will be very different compared to those of the past. While historical knowledge will be useful, finding solutions to the water problems of the future will require additional skills and new mindsets. It will also require a determined attempt to coordinate energy, food, environment, and economic policies of a nation, all of which have intimate linkages to water. Each will affect the other and, in turn, be affected by the others. Policies in all these areas will be also influenced by exogenous forces, such as demography, advances in technology and information, climatic change, free trade, and increasing social activism. Resultantly, water management in South Asia will change more during the next 20-30 years than ever before. Water problems can be solved. South Asian countries are not facing a water crisis because of a physical scarcity of the resource but because of poor management. Awareness on the need for societal adaptation to hydro-climatic constraints, and strong enough institutions to deal with them must be developed. Land/water linkages to the ecosystem will have to be integrated, and a catchment-based land/water/ ecosystem approach needs to be developed. The goal has to meet both societal needs and environmental sustainability conditions.

Conceptually, and at a policy level, 'security' has acquired multiple connotations and is increasingly being perceived as issue-based rather than as an overarching idea. Notwithstanding the critics who regard broadening the security ambit as leading to intellectual incoherence and as being counterproductive to devising solutions, modern security concepts are coming to terms with multiple meanings. With water becoming an increasingly challenging resource, its salience on national security has assumed enormous significance. Water security implies affordable access to clean water for agricultural, industrial and household usage and is, thus, an important component of human security. Along with food and energy, water forms a critical part of the 'new security agenda' and redefines the understanding of security as a basis for policy response and long-term planning. The 'securitisation move' of an existential issue such as water generates political attention, public awareness, and policy-initiatives. There is, however, the risk that the issue can become vulnerable to political vested interests and linkage politics, with the possibility of solutions being manipulated within the political context.

When examining water issues through the lens of security, it is important to understand water as a strategic commodity – one which is vulnerable to pressures of availability and demand. Equally important is the fact that water is a function of the relationships among social, political, and economic factors. These fundamentals reiterate the looming water crisis as a multiple challenge, encompassing not only the users and their beliefs and attitudes but also the politics and policies that determine its distribution and utilisation. It is a principal preoccupation of states now to grapple with the coming water scarcity, and prioritise it as the most important issue.[1] Since scarcity issues are all-encompassing, technological over-reliance helps alleviate the immediate effect of water shortage. However, the long-term approach to the reality of a world increasingly water-stressed would require greater response to address the political and institutional problems. While the likelihood of tension and conflict emanating from the consumption and distribution pattern of water resources cannot be underestimated, historically, such resources have been used more as the means or rationalisation of conflict than seen as its cause.

Three factors contribute to water resource being a scarcity threat: depletion and degradation; increased demand, and uneven distribution. Continued population growth and global warming, along with over-consumption, inadequate conservation and wastage, will further put pressure on water resources. This suggests that there is a considerable need for resource planning, regional cooperation to manage disputes, and 'adaptive governance'.[2]

Water Issues: Conflict or Cooperation?

As argued in Chapter 1, freshwater is a *precious* resource and its *possession* bestows power. The 'preciousness' and 'possession' in geopolitical mechanics renders water a strategic commodity, and its role as a strategic asset or vulnerability (in terms of supply and demand) cannot be underestimated.

In the geopolitical framework water is taken as a 'good', and conceptualised under the model of resource scarcity.[3] The 'geo-politicisation of water' is associated with the 'instrumentalisation of water' and, therefore, the common usage of the term 'water wars'.[4] Thus, water becomes a resource[5] of contention, and conflict is generally reduced to the question of who has the 'good' and how much, who needs and how much (or how much is needed) and subsequently, what the affordable cost of 'procurement' of such a 'good' would be in economic, political, or military terms. The preceding chapters have investigated why and when states choose to cooperate over water, or why and when states tend to use water as a 'bargaining tool' and an 'instrument of politics'.

'Water in a Changing World'[6] will assume greater salience and, as it does so, the drivers impacting water resources – whether climate variability and security issues or electricity-generation and migration – will need to be factored in and solutions searched for. A considerable amount of technical and scientific knowledge developed in recent years points towards the potential of water scarcity becoming a key driver of tension and conflict within societies and states. The possibility of inter-state wars arising from water-related issues has been much talked and written about.[7] However, one can dispute such an alarmist prognosis, and treat water wars more as fantasy than reality. History tells us that the only recorded water war was some 4,500 years ago, when the two Mesopotamian city-states, Lagash and Umma, went to war. History also shows that, between 805 AD till now, countries have signed more than 3,600 water-related treaties. Thus, there seems to be more active cooperation over water than actual war.

Security practitioners need to take water issues into account as part of their arsenal of tools, and explore two primary questions: What role do water issues play in stimulating international conflict and cooperation? Are conflicts over water sharing likely to be more 'within' (intra-state) or 'between' states (inter-state)? The divide in terms of scope and focus is of obvious policy importance, particularly since threats emanating from water scarcity feature regularly in policy reports.

Rivers and Riparian Relations

Rivers constitute a significant share of the world's available supply of fresh water, and are the most widely shared resource on the planet. There are more than 260 river basins that are shared by two or more states, and there are 145 treaties in existence today.[8] Rivers with their distribution, utilisation, and the potential for hydroelectric power form a critical component of inter-state relations. It is commonly acknowledged that, in spite of non-binding international law and rules for managing river water basins, treaties serve as the best management tool. A large number of riparian treaties reinforce an argument that river waters are a 'catalyst' for cooperation even among hostile states than an 'inducement' for conflict. In the past 50 years, there have been only 37 cross-border disputes which have involved violence, while numerous initiatives on water-sharing have been negotiated and signed.[9]

The hydrologic reality strongly dictates that hydro-cooperation is the best possible means of optimising trans-boundary river waters. Even the 1997 UN Convention in spirit speaks of an 'equitable utilisation of water resources' and 'meeting vital human needs'.[10]

The ability to cooperate on trans-boundary rivers limits the 'geopoliticisation' and 'instrumentalisation' of water. Crucially, it helps to focus on key issues such as river basin management, equitable distribution, as well as the assessment and monitoring the potential impact of climate change on the region's water resources.

The basis for any river water treaty is to continuously find an equitable approach for meeting vital human needs. Water treaties, particularly in regions where scarcity is high, are also a barometer to gauge state behaviour and the political climate. This raises a few interesting questions: To what level does the changing political climate affect existing treaties? Do signing of river water treaties lead to more cooperative ventures between the riparians concerned, and thereby enhance the overall peace environment in the region? Is the negotiation process preceding the signing of a treaty a final solution? Or is it only a provision that temporarily conceals the claims and counter claims as well as the real and perceived fears of the riparians (particularly the lower riparian)? Does 'the real and perceived fears' lead to non-compliance of the treaty along with an overriding 'militarised' approach in which the 'possession' of water is determined unilaterally? And finally, what are the linkages associated with trans-boundary waters?

India and its Riparian Relations

The water challenge before India in the coming years will be two-dimensional: to manage its water resources better; and simultaneously, to manage better its riparian relations with its neighbours. For India, as an active regional player, water management will be crucial to conflict management. Being international, indispensable, and emotional, water can serve as a cornerstone for confidence building as well as a potential entry point for peace. Thus, India will have to balance its growing water needs and larger security concerns with effective hydro-diplomacy. The salience of rivers in India's relations with its neighbours is bound to increase in the coming years.

Rivers have many uses. Some are consumptive in nature and some non-consumptive. The non-consumptive uses (such as navigation and hydroelectricity generation) are less problem generating than non-consumptive uses (such as drinking water and water for irrigation). Because river uses are so intricately linked, riparian treaties seldom specify the ultimate use of river waters.

No doubt, the hydrological profile of the Subcontinent – the Indus basin and the GBM – link countries together; but they also bitterly divide them. With population pressures and the need to achieve developmental goals, disputes and grievances arise over the use of and control over the rivers. Structures like dams and barrages create upper-lower riparian tensions that can be the potential cause of conflict. Numerous bilateral treaties exist; but they are often hostage to prevailing political animosities. Resource nationalism will increasingly dominate the hydrological contours of South Asia, and will largely define regional politics. Many of the existing riparian treaties will come under pressure over the sharing and harnessing of river waters.

India's riparian relation will become progressively fragile with Pakistan, Bangladesh, and Nepal which are continuously raising concerns over the regulation and sharing of river waters. China's aggressive south-to-north water diversion projects on the rivers that originate from Tibet, particularly on the Yarlung, are opening up a new front of uncertainty in Sino-Indian relations.[11]

The friction in bilateral relations will increase if mutually acceptable bilateral or multilateral frameworks for cooperation to deal with integrated development of water resources are not effectively re-worked. Other externalities – like the development of satellite technology that will enhance the ability of states to chart flow volumes and give real-time data on water uses – will result in heightened public awareness and contribute significantly to enforcing river water allocation. In such situations, many of the existing treaties will have to be evaluated afresh, and many new treaties based on new hydrological knowledge will need to be framed. The geographical contours of India both as an upper riparian and a lower riparian will become the epicentre of the new riparian politics and diplomacy over trans-boundary rivers.

The crux of the problem over trans-boundary rivers is that this resource is neither seen exclusively as a public good (defined as non-rival and non-excludable) or a private good (defined as rival and excludable).[12] With no agreed definitional demarcation, trans-boundary rivers are often viewed as 'collective goods' or 'common pool resources'. But to expect countries to accept that sovereignty has to be exercised collectively, particularly in respect of the global commons, is wishful thinking. The treatment of rivers as a 'good' in the Subcontinent will primarily be interpreted within the regional asymmetric/symmetric power configuration. The upstream-downstream supply disputes will be a common feature of riparian politics. Moreover, international laws on allocating water within river basins are difficult to implement, and often contradictory. The UN Convention on the Non-Navigational Uses of International Watercourses – approved in 1997 by a vote of 104-3 – was ratified in 2014, and requires watercourse nations (Article 5) to participate in the use, development, and protection of an international watercourse in an equitable and reasonable manner.[13]

In spite the complexities and potential deadlocks of river water sharing, riparian treaties are a common feature in the Subcontinent. As middle riparian, India is at the centre of hydrological relations. The existing treaties underline an important element of river discourse: that while there are factors that hinder formalised river cooperation, there are equally countervailing factors that enable peaceful sharing. In its difficult neighbourhood, India will have to live with not only a trans-boundary rivers arrangement but also critically consider the scarcity problem based on strategic rationality, hydrological effectiveness, and economic viability in order to reshape the existing treaties.

Thus, the uneasy marriage of politics and external drivers that govern most shared waters cannot be overlooked. The challenge for decision-makers in the midst of palpable tensions and strife over trans-boundary waters (inter-state) and inter-provincial water transfers (intra-state) is to constantly find new mechanisms and approaches to reduce the tension resulting from water issues. India requires a new and integrated framework to deal with water security issues. The existing national water policy is a stand-alone document which does not integrate with the country's food, energy, and health policies. Nor does it take into account the impact of climate change. Also, riparian relations have not been factored in. Thus, the security discourse in the region cannot be complete without discussing water security.

What does India as a Middle Riparian Do?[14]

India's middle riparian position has lacked the emphasis both in the neighbourhood as well as it with China. India is downstream with China and thus has concerns over water release, planed diversion and dams. On the other hand, it is upstream with Bangladesh and Pakistan and has responsibilities towards water sharing. How should India then approach this riparian configuration?

India's National Water Policy (2012) was finalised and adopted by the National Water Resources Council on 9 August, 2012. Some of the salient features are: "addressing issues such as the scarcity of water, inequities in its distribution and the lack of a unified perspective in planning, management and use of water resources."[15] The NWP with its overarching general principles on water prompts states to draft their own water legislation. Some of the basic principles that govern the NWP are:[16]

- The principle of equity and social justice in the use and allocation of water.
- An integrative approach to planning and management of water resources that takes into consideration local, regional and national contexts with a strong emphasis on the ecosystem protection.
- The economic value of water to promote its conservation and efficient use after basic needs are met.
- The river basin as the basic hydrological unit.
- Stress on new knowledge, improved data collection and assessment along with improved water management methodologies and conservation technologies will generate a more comprehensive water profile in terms of quality, quantity and distribution.
- To incorporate transboundary aspects of water sharing into the National Water Policy. Therefore a focus on water cooperation with India's neighbours and a foreign policy priority.

- Attention to common interests in the region, like water management, through multilateral efforts involving river-basin actors.
- Water security in the region based on rationality of water, prudent national water management and sensible co-riparian relations.

With Pakistan: The Indus Waters Treaty (1960) was a pragmatic Treaty that took into consideration the concerns of Pakistan. That it remains functional is a tribute to the framers and negotiators. The Treaty is regarded as a success story of India-Pakistan relations. However, as a lower riparian, Pakistan has used it as a tool to garner international sympathy. Fears persist that water could be a catalyst for conflict. Further, China's upstream requirement and consequent structural designs on the Indus will prompt India to construct storages on the western rivers and guarantee for itself additional water.

- The Treaty provides for India a storage capacity of 3.6 MAF, which it has not achieved so far. India's grievance or complaint of the Treaty is that the stringent provisions thwart India from constructing hydro-project works on the western rivers. Thus, a 'modification' of the provisions of the Treaty might suit India whether through renegotiations or through establishing an Indus II Treaty. In a re-negotiation, there are dangers that India, as an upstream, might be forced to consider giving away more water Pakistan.
- India's upstream position can be a counter-weight to Pakistan-sponsored terrorism by calling for a revision or even unilaterally abrogating the Treaty. This is permissible in international law.
- In the meanwhile, India should take all steps to utilise the waters of the eastern rivers which are going to Pakistan even though India has complete rights over them. Likewise, India should build projects on the western rivers to harness their power potential while keeping within the bounds of the Treaty.
- In addition to surface water issues, ground water abstraction is a matter of serious concern. Pakistan and India share a continuous water aquifer which is not clearly demarcated between the two countries. Over-abstraction of groundwater in Punjab (Pakistan and India) is affecting the aquifer's water quality and quantity. At one level, there is hardly any bilateral engagement on the aquifer and the knowledge base in terms of data and studies are insufficient to know the current rate of abstraction.

With China: The Hindu Kush Himalayas (HKH) are the 'water tower' of South Asia and Southeast Asia. It covers an area of more than 4 million sq km and includes eight dependent countries (Afghanistan, Pakistan, India, Nepal, China, Bhutan, Bangladesh and Myanmar. The snow, ice, permafrost and rainfall of the HKH are the sources for the ten largest river basins in Asia including the Indus River System and the Ganga-Brahamaputra-Meghna basins. The significance of Tibet as part of the HKH cannot be overlooked. China annexed Tibet in 1950, and thereby gained control over the vast glaciers and the headwaters of the rivers. Beijing has strengthened its political and economic control over Tibet wherein India and China have a complex, unresolved boundary dispute. China as an upstream controls the headwaters of the Yarlung (Brahmaputra) and the Sutlej which flow into India as also the Indus that flows through Ladakh before it enters Baltistan in POK. Thus, water will assume higher priority in Sino-Indian relations.

- Tibet's rich biodiversity and diverse ecosystem is vulnerable to climate change. China is well within its sovereign right to utilise and divert the waters of the rivers in the Tibet region in order to meet high demand in its arid north. But should China be the lone stakeholder to the fate of the ecosystem in Tibet? What will happen in the downstream nations that depend on the rivers originating in the plateau?
- India and China have in 2013-14 enhanced their MoU on the hydrological data exchange on the Yarlung/Brahmaputra as well as the Sutlej. India should continue to pursue a more broad-based level of hydrological dialogue with China.
- India as downstream with China should discuss its concerns with other downstream countries like Bangladesh. A collective downstream response and water dialogue with China should be based on norms of equitable utilisation, 'no harm' policies and ecosystem protection in Tibet. Regional attention to defining water as 'commons' would go a long way toward preserving and sharing the waters of Tibet.[17]
- Another option for India should be to take the initiative to propose tapping the U-bend from Tibet to Assam as a major regional carbon-saving project with international collaboration and the basis for a South Asian-China-ADB project.

- Transboundary water management is important to the sustainable development of the Yarlung/Brahmaputra basin countries. China, India, Bangladesh and Bhutan should enhance communication between NGOs and CSOs in water resources distribution, integrated river basin management and flood control on the Yarlung/ Brahmaputra river basin.

With Bangladesh: The Ganga Treaty has been generous towards Bangladesh in terms of water sharing. However, as is typical of a lower riparian, Bangladesh continues to complain against the treaty and demand more water. Recently, the sharing of the Teesta waters, the construction of Tipaimukh dam, and the Indian project for interlinking of the waters have caused much concern. While it is necessary for India to continue its dialogue with Bangladesh proposing joint river basins wherever possible, it is equally important for India to guard its own interests. Bangladesh cannot change its lower riparian position and, thus, has to accept cooperative arrangements based on water sharing, and not on water rights. Nearly 54 rivers flow from India into Bangladesh. Given India's advantage as an upper riparian, it should use its hydrological position to leverage other important interests in particular security issues.

- The Ganga Treaty between India and Bangladesh is an agreement to share surface waters at the *Farakka* Barrage. But the treaty divides the water flow without sharing the value and uses of the river between the two countries. It should take into account a whole-of-basin approach to river management with the integration of grassroots voice particularly the poor women and men of those location.
- Instruments such as the UN Watercourses Convention of 1997 should become a guiding light in improving the standard of living of the largest concentration of the poor people in the world on the Ganga basin.

With Nepal: India and Nepal have had a long history of wide-ranging cooperation on water issues; but this has not been free of problems. India is a lower riparian but does not fear being either denied water or being flooded. Many of the joint projects relating to flood control, irrigation, and hydroelectricity have been myopic and mismanaged. There are a number of existing bilateral mechanisms which are devoted to different aspects of water cooperation. Fluctuating political relations have deterred water resources development. Besides other factors, Nepal's mistrust has been reinforced by what it perceives to be unequal treaties – starting from the Sharada Dam construction (1927), Kosi Agreement (1954), Gandak Agreement ((1959), Tanakpur Agreement (1991), and the Mahakali Treaty (1996). The efficacy of bilateral cooperation needs to be increased by improving the working of the existing bilateral mechanisms, including the Joint Committee on Water Resources. India's efforts should be to address Nepal's concerns in as reasonable manner, and to provide it adequate financial and technical assistance.

- The first step towards this is to rethink the entire riparian approach with Nepal which has largely been based on hydro-generation. Moreover, seismic factors make the terrain unsuitable for building large dams and large storages. India's focus should be on flood management and control; the prevention of sediment, inundation and soil erosion; and irrigation benefits for both.
- While reconstructing a new and trustworthy relationship should be high on the agenda, India can simultaneously think of identification and feasibility studies on small, medium and, if required, big dams should be undertaken; building up confidence in shared benefits, and medium-size hydroelectric projects can also be initiated.
- The Kosi and Gandak treaties should be revisited, and the positive elements should be repackaged. New hydrological knowledge and new methods of river water management should be wholeheartedly introduced in framing future India-Nepal water cooperation policies.
- With the current NDA government's neighbourhood thrust, Nepal is going to build two major hydropower plants with USD 2.4 billion in Indian funding. It includes a 900MW, USD 1.4 billion plant in Upper Karnali to be built by GMR, making it Nepal's largest ever foreign direct investment and another 900 MW USD 1 billion Arun III hydroelectricity plant in northeastern Nepal. Both these projects are under a "build, own, operate and transfer" contract. Nepal gets an equity stake in the Upper Karnali plant and access to 12 per cent of the electricity output for free. Ownership of the plant will be transferred to Nepal after 25 years. For Arun III, Nepal will get 21.9 per cent of the electricity for free but no equity. That plant will also be handed to Nepal after 25 years.

A growing need to synchronise internal water management measures with external riparian policies is critical. Although India has low per capita water consumption, it lags in the efficient use of water across sectors. A policy revamp, which moves away from a narrowly understood framework of 'water management' to a broad-based and wide-reaching 'water resource management'[18] is the need of the hour. This would require treating river systems, particularly the Ganga-Brahamaputra-Meghna (GBM) and the Indus, in a holistic way and reorienting hydro-diplomacy on a multilateral basis rather than just a bilateral format.[19]This would entail a shift from 'sharing waters' to 'sharing benefits'. The GBM and Indus basins account for two-thirds of India's water potential. Further, any water outlook will necessitate interdisciplinary approaches, linking together natural sciences, politics and policy. The challenge for India will be to imbibe hydro-diplomacy in it's over all regional diplomacy – a not so easy task as India's diplomacy has traditionally been bilateral rather than multilateral.

And Finally...

A Common Basin Management or Comprehensive Basin Management is now being considered as the new confidence building measure on the GBM that includes Bangladesh, Bhutan and Nepal. After the Amazon and the Congo, the GBM is the third largest freshwater outlet into the ocean. The GBM is the most densely populated (630 million), with the largest number of poor people in the world. Thus, it is critical to transform water resources into sustainable economic development. The riverine neighbourhood would require an integrated development plan wherein water resources needs to developed and managed in a rational, efficient and equitable way.

A number of studies have demonstrated that integrated water management in the GBM basins can offer greater benefits than those that can be achieved through isolated national efforts. Having basin coverage of 64 per cent of the 1.7 million sq km of the GBM, India has to take the lead and the initiatives. Some common issues and challenges are:

- Wide variations between 'peak' and 'lean' flows
- Future climate change impacts may aggravate the situation further
- GBM is in the monsoon region, in which 80-90 per cent rainfall is during 4-5 months of the South-West monsoon
- The rivers carry a large amount of sediments which affects the

carrying as well as retention capacity of the rivers, compounding the adverse effects of floods

- All the countries in the GBM have national water policies. Each needs to be studied carefully as the emphasis is different. These are; Indian Water Policy 2002 of the Ministry of Water Resources; National Water Policy of Bangladesh, 1999; National water Plan of Nepal, 2005; National Water Policy of Bhutan 2003.
- The challenges facing the GBM include lean period sharing; augmentation of flow in the lean period; hydropower generation and distribution; cooperation in flood management; water quality problem and protecting ground water.
- National water policies of the GBM basin countries are in conflict with each other, leading to misunderstanding and dispute. This fails to bring in consensus among the riparian.
- Efforts on enhancing navigation

However, there are many opportunities that also exist. These are:

- India has an advantage to prioritise basin management. Regional cooperation on water resources is recognised in the water policy of Bangladesh, with exchange and sharing of information and data, and joint assessment of the basins' potentials. Nepal too recognises the potential for sharing water resources 'benefits' on equitable terms, and seeks to enhance regional cooperation in sharing and exchange of data, and improving disaster forecasting and warning systems. The water policy of Bhutan expresses similar pledges with regards to regional cooperation.
- Regional diplomacy should emphasise sharing benefits through an integrative approach; learning from – and possibly involving – successful institutions like the UNESCO and the UNEP in the basin management approach.
- Basin management will require both structural and non-structural approaches

Structural approaches would include:

- selection of suitable multipurpose storage sites in Nepal;
- the prospects of constructing reservoirs in the middle and lower Ganga basin in India to store monsoon flows. The reservoirs are

likely to mitigate floods in the downstream reaches of the Ganga as well as make the monsoon water stored available for dry season augmentation of flows, thus increasing dry season irrigation potential and also the possibility of river navigation.

- an integrated plan for hydroelectric development and sharing through an interconnected grid across the borders.

Non-structural approaches would include:

- Sharing or exchange of data and information.
- Joint studies and assessment, such as the Bangladesh–Nepal Joint Study Team (1989) that identified and recommended 30 potential reservoir sites in Nepal. Joint studies on glaciers and GLOF (Glacial Lakes Outburst Flooding)
- Measuring and compiling hydrological data from the number of tributaries that join the GBM. Tributaries flowing from Nepal contribute the major flows of the Ganga (about 40 per cent of the annual flow and 70 per cent of the dry season flow).
- Fusing the national water policies of the GBM countries into one regional document.
- Coordination and integration within different ministries of different countries – that is, Ministry of External Affairs, Ministry of Environment, and Ministry of Water Resources.
- Emphasis on inland river navigation which is not only cost effective but also energy efficient. The regulation of flow through the creation of multi-purpose reservoirs in the upper reaches of the rivers will open opportunities for inland river navigation in the downstream reaches. All the countries in the GBM could benefit. The Ganga is National Waterway 1, and the Inland Waterways Authority of India (IWAI) needs to be cooperated and strengthened in basin diplomacy. Surveys of the river bed, determining the deepest part of the channel over which boats can safely travel, and mapping this information will be a critical role of the IWAI. Also, this data and maps should be easily available to transporters. Smart cities on the Ganga will require smart ports, factories, and markets around the ports.

It will also be necessary to remain engaged with China not only on the Brahmaputra but also on the many tributaries of the Ganga which originates from the Tibetan plateau. During the visit of Chinese Premier Li Keqiang to India in May 2013, serious time and discussion was given to water issues. India's proposal of a joint mechanism for better transparency on the dams being constructed on the Yarlung was diplomatically appreciated.

Trans-boundary water is an important feature of South Asia. Riparian cooperation is desirable in conformity with existing agreements. Further cooperation on rivers will also require relevant arrangements that take into account the interests of all the riparian states concerned. Looking at international rivers for good practices and right lessons will be important in terms of shared knowledge on rivers management. A new rationality of hydro-solidarity will have to be developed between those living upstream and those living downstream in a river basin.

NOTES

1. For example, the UPA government's Common Minimum Programme, the US intelligence community overview of 'Global Trends 2015', and the UN Report on 'Our Shared Responsibility'. For the complete text of the UPA's Common Minimum Programme, see *Strategic Digest*, 34(9), 2004; for *Global Trends 2015,* see http://www.cia.gov/cia/reports/globaltrends2015/globaltrends2015.pdf, or for a summary, see http://www.cia.gov/cia/reports/globaltrends2015/. For the complete text of *A more secure world: Our shared responsibility, Report of the Secretary General's High-level Panel on Threats, Challenge and Change*, see http://www.un.org/secureworld/report2.pdf
2. While 'adaptive management' is an approach to the conservation and utilisation of natural resources (like water), it seeks, through experimentation, to maintain ecologically sustainable levels of use. 'Adaptive governance' involves political choices, that is, specification of policy objectives, allocation of revenue, imposition of regulatory controls, and the allocation of gains and losses necessary to achieve political equilibrium regarding levels of water quantity and quality, whether these are ecological sustainable or not. See, Lawrence Susskind, 'Adaptive Governance", in John T Scholz and Bruce Stiftel (eds.) *Adaptive Governance and Water Conflict*, Washington DC: RFF Press, 2005, p. 142.
3. The scarcity-conflict model is fast becoming conventional wisdom in foreign policy, population and environment circles. It has been structured by Stephan Libiszewski and Homer-Dixon and popularised and sensationalised by writers like Michael Renner, in 'Ending Violent Conflict', *Worldwatch Paper* 146, 1999, and Robert Kaplan, in 'The Coming Anarchy', *Atlantic Monthly*, February 1994, pp. 44-76. Kaplan proclaimed the environment was the most important national security issue of the 21st century.
4. 'Water wars' is a much hyped alliteration. The prediction of water wars seems to be sensationalist and alarmist.
5. If water is seen as a 'source' (a source of life, and without which nothing survives), then the entire perception of water changes from being one of hostility, to one of cooperation and sharing.
6. The theme was introduced in the World Water Forum in Istanbul, March 2009
7. In the early 1980s, the Egyptian Minister of State for Foreign affairs Boutros-Boutros Ghalisaid, 'The next war in our region will be over the waters of the Nile.' In 1991, a

few months before being appointed as the Secretary General of the United Nations, he reiterated, '...the next war in the Middle East will be fought over water, not politics.' Thereon, 'water wars' as a dramatic alliteration was used in the article by Joyce Starr. In 1995, World Bank Vice-President Ismail Serageldin made a much-quoted prediction about the future of war: 'If the wars of this century were fought over oil, the wars of the next century will be fought over water.'

8. Of the 145 treaties, 124 are bilateral; 21 are multilateral of which 2 are drafts. Such facts are found in the databank compiled by the Department of Geography and the Center for Freshwater Studies, Oregon State University in conjunction with the World Bank and the US Institute of Peace. The databank catalogues the treaties by basin, countries involved, date signed, treaty topic, allocations measure, conflict resolution mechanisms, and non-water linkages. For a detailed assessment and comparative analysis see: Jesse H Hamner and Aaron T Wolf, 'Pattern in International Water Resource Treaties: The Trans-boundary Freshwater Dispute Database', *Colorado Journal of International Environmental Law and Policy,* 1997 Yearbook.

9. Cited in 'UN World Water Development Report', 2004

10. The '1997 UN Convention: Law of the Non-Navigational Uses of International Water Courses', drawn from the earlier Helsinki Rules of 1976, is the main basis for international water law. It requires the ratification of 35 nations before it comes into force. Only a few countries have signed and ratified it. Though not binding, the 1997 Convention serves as a 'guiding principle', which casts its influence in negotiations on river waters. The 1997 Convention lays out two significant points that are the heart of any negations on water courses:
    1. 'To assure equitable utilisation of water resources among all riparians on an international water course, with priority given to meeting vital human needs, and
    2. The obligation not to cause significant or appreciable harm to other riparian states."

    For the full text, please see, http://untreaty.un.org/ilc/texts/instruments/english/conventions/8_3_1997.pdf

    The keyword 'equitable utilisation' as a principle gives no claim for an upper riparian state to have 'sole and exclusive' rights over the use of water derived from sources within their territory. For example, the Syrian and Iraqi (down-stream countries) claims for use of water from the Tigris and Euphrates rivers that originate from up-stream Turkey is accepted under the 1997 UN Convention. Likewise, Syria's claim for absolute rights to the waters of the Jordan River without taking into consideration the downstream claims of Israel is also unacceptable. The 1997 Convention does recognise the 'legitimacy of prior and historic user rights but the claim that historic use assures immutable and sole water rights is also not absolute in terms international law.' See Hillel Shuval and Hassan Dweik (eds.), *Water Resources in the Middle East,* Berlin Heidelberg: Springer, 2007, pp.10-11

11. The Yarlung-Tsangpo becomes the Siang in Arunachal Pradesh. This, in turn, joins the Luhit, Dibang and Noa Dihing near Sadiya to form the Brahamaputra.

12. Jaroslav Tir and John T Ackerman, 'Politics of Formalised River Cooperation', *Journal of Peace Research,* vol.46, no.5, September 2009, p.623

13. The document requires ratification by 35 countries to enter force. As of 2008, only 16 countries had ratified the convention. Article 7 of the Convention entitled 'Obligation not to cause significant Harm' is particularly controversial. It would require states, 'in utilising an international watercourse in their territories...take all appropriate measures to

prevent the causing of significant harm to other watercourse states and compensate sharing states from any such harm.' Legal experts argue that a confliction is imminent since a state may have legitimate uses for a watercourse in its nation that can negatively impact other nations.

14. These recommendations are an updated and modified version of the IDSA Task Force Report on India's External Water Security Challenges, which the author had the privilege to be the lead writer.
15. Report Summary, Draft Water National Policy, 2012, PRS Legislative Research, June 2012, at http://www.prsindia.org/parliamenttrack/report-summaries/summary-on-draft-national-water-policy-2012-2431/
16. See National Water Policy, 2012, Government of India, at http://ecnphlgnajanjnkcmbpancdjoidceilk/http://wrmin.nic.in/writereaddata/NationalWaterPolicy/NWP2012Eng6495132651.pdf. Also see, Summary Records of Proceedings of 6th Meeting of National Water Resources Council on 28 December, 2012, at http://ecnphlgnajanjnkcmbpancdjoidceilk/http://wrmin.nic.in/writereaddata/NationalWaterPolicy/SummaryRecordofProceedings6576369201.pdf
17. Uttam Kumar Sinha, 'Tibet's Watershed Challenge', *The Washington Post*, 14 June 2010
18. 'Water management' has traditionally been understood as the manipulation of water for specific uses, leading to independent legal regimes and the development of isolated water-based projects. The concept of 'water resources management' includes the comprehensive protection, development, and utilisation of the whole of a given body of water, surface and underground, constituting one single hydrologic unit. Water resource management is viewed as being 'rational' and 'integrated'. It embraces the 'conjunctive use' of surface and underground water resources, the control of their quality as well as harmful effects; their sustainable development for multi-purpose beneficial utilisation and the administration of corresponding use rights. See, Bernard J. Wohlwend, 'Equitable Utilisation and the Allocation of Water Rights to Shared Water Resources', at http://www.aida-waterlaw.org/PDF/EQUITABLE.pdf
19. See, Jayanta Bandyopadhyay and Nilanjan Ghosh, 'Holistic Engineering and Hydro-Diplomacy in the Ganges-Brahmaputra-Meghna Basin', *EPW,* Vol. 44, No.45, Nov. 7-13, 2009, pp.50-60.

Index